No Going Back – Journey to Mother's Garden has sold more than 31,000 copies and has been translated into Dutch, Portuguese and Catalan. Martin Kirby's last book, *Count The Petals Of The Moon Daisy*, a novel that goes to the roots of countless English Americans, is now a feature film project.

Martin, Maggie Whitman, and children Ella and Joe Joe, were seen by millions around the world in the very first episode of the Channel 4 TV series: ***'No Going Back'.***

Mother's Garden, in the mountains of The Priorat, Catalonia, has been their wild side home for ten years.

By the same author

Albion (Jarrolds) 1998
ISBN 978 0711 710 27 6

No Going Back – Journey to Mother's Garden
(Times Warner), 2003
ISBN 978 0751 535 48 8

Count the Petals of the Moon Daisy (Pegasus), 2007
ISBN 978 190 3490 29 7

Shaking the Tree
Mother's Garden – The Growing Years

For Joe and Lorraine Williams
of Imagine Mozambique

And my Maggie
The making of me
I love you

Martin Kirby

Shaking the Tree
Mother's Garden – The Growing Years

Pegasus

PEGASUS PAPERBACK

© Copyright 2010
Martin Kirby

A CIP catalogue record for this title is
available from the British Library

ISBN-978 1903 490 59 4

*Pegasus is an imprint of
Pegasus Elliot MacKenzie Publishers Ltd.*
www.pegasuspublishers.com

First published in December 2010

**Pegasus
Sheraton House Castle Park
Cambridge CB3 0AX England**

Printed & Bound in Great Britain

With loving hugs for our children Ella Kirby, Joe Joe Kirby, our incredible families and our wonderful friends.

And special thanks to Heather Tamplin, readers of the Eastern Daily Press and Yorkshire Post, and the thousands of people around the world who have visited Mother's Garden, stayed at the cottage, bought our olive oil or just sent good wishes to us over the years.

You make all possible…

mother's garden

UP THE GARDEN PATH

A prologue

*This was among my prayers: a piece of land
not so very large, where a garden should be
and a spring of ever-flowing water near the
house, and a bit of woodland as well as
these.*

Horace (65BC-8BC)

Thank you, Horace. Timelessly astute. That should touch a nerve or two (million).

Hands up all of you who are now sighing wistfully at the thought of such pastoral simplicity far from the madding crowd?

I did.

I sighed; then I yearned until there was no stopping me. Maggie, my partner, shared that feeling truly, madly, deeply. At the end of our thirties we seriously wanted something simpler, something else. Well, it's not too much to ask, is it? A little, space, peace, time, harmony, nature, sense of adventure – oh, and fun – away from all stress, overload and constant running without getting anywhere that had amounted to life thus far.

No – it was more than that: A tilt while we still could, a search for more colour and contrast, a ripping yarn to tell grandchildren, an experience to plug the yawning gap in the meaning of a life that is whizzing by, to do something radically different in the perplexing quest for fulfilment. (Dwell on the word fulfilment for a moment. Scary, isn't it?)

If you know what Eddie Waring, Napoleon Solo, Pan's People, Big Daddy and Bernie the Bolt did for a living (not collectively) then you, like me, are of an age when the questions in your head come faster than the answers. I'm a 1958 baby, raised and then released into a golden age of cotton wool middle England, too young to understand the missile crisis and Kennedy shock, but so transfixed by the Apollo 11 moon landing and the 1970 Leeds vs Chelsea FA Cup final and gruesome replay that the names of those involved are tattooed on my brain. I can hear Neil Armstrong's voice clear as a bell and still see Eddie Gray tormenting David Webb. Then came the hairy seventies. I grew untidy and upwards to a gangly 5ft 11 and three quarter inches, scraped some qualifications, got to grips with the opposite sex, bought a Ford Capri, somehow collared a reasonable career and then dabbled with the tricky task of appearing adult (er, signing up for a 100 per cent endowment mortgage because the aptly titled broker said I'd make a pretty penny; living permanently in the red; pretending I understood the lyrics to the Yes album Tales from Topographic Oceans; trying to grow a moustache; reckoning that voting for Jeremy Thorpe's Liberal Party made me interesting).

"You've got your whole life ahead of you boy," people would say. And when they stopped, I still thought it. Thatcher was deposed. I clocked thirty. I had a yolk yellow 3-door 1.4cc Volkswagen Golf, tidy hair, wore a suit to the office, listened to James Taylor in the bath and kept up the mantra. Still loads of time, Laddie.

Then – shzam – a decade evaporated and I was suddenly fast approaching an age I once thought ludicrously fuddy-duddy. The brevity of life welled every time I saw a scrawny, baggy-eyed photograph of myself, namely all of them. In my case that dovetailed with receding hair regrouping in my nose, flatulence failing to be funny anymore, my toenails turning to ivory, and a serious improvement in Radio Two. Well it's true. I would rib my mum mercilessly about *Sing Something Simple* and her inability to get into the Lou Reed groove, but you should hear the rubbish on Radio One these days.

You, Kirby, are going to die someday seriously hence, a voice between my enlarging earlobes kept goading, so time to start questioning what you have done of worth and, more pertinently, what you want to do before your meter runs out. Or, rather, what you don't want to do any more. Those whose role it was to make me feel young – parents, the older generation – were departing this earth, or had gone and bequeathed me a mahogany commode. And this patchwork, ill-fitting clown's trousers of a human race really – REALLY – wasn't funny anymore.

What little I see these days of the higher-octane nations like the one I left, it seems nothing has changed. There's just no let-up. It's exhausting. Everyone's wound up; the news coming from all angles is bleak beyond words then – oh joy – you are told that, due to calamitous cock-ups and unbelievable gluttony in the banking world, it appears everyone's existence is dependent to a terrifying degree on incomprehensible digital financial fictions. Lob in a few humongously greedy executive berks, several ghastly Budgets and large lumps of seething envy, simmer for a decadent decade, then watch as it comes to a lava-like boil.

At times like these, folk are inclined to let rip, rethink, maybe even make BIG changes.

So, had enough? Want out? Are you storing up for when the economy gets off the ropes some serious or idle dreams of bidding Britain au revoir, adios, arrivederci, αντίο (Greek) or even довиждане (Bulgarian)? Was the 2009 UN report saying Britain had slipped out of the top twenty in the quality of life league, despite there being more mobile phones than people, painfully near the knuckle?

I have been absent for ten years, although my wife will tell you that, cerebrally, it's considerably longer. Allow me, then, a lapsed but not excommunicated Briton, a lanky north European with a Viking name attempting to live the Latin good life, to lead you up my decidedly rocky Mediterranean garden path. No rose tints, just a battered old cap to shield the eyes.

We'll weave through undefeatable brambles and abandoned hazels to the top of our little Catalonian farm, where the boar plough up the soil with their noses and the round-shouldered buzzard waits on the power line for lunch. Once there, we can sit and slap our necks under an olive tree beside the breathtakingly bio-diverse, weed-infested vineyard. We can stare at the mountains, worry about global warming, economics and stinger ants in our pants, fret about forest fires and pray for rain, and I can tell you some more home truths about making a new life that many think is a quantum leap from Coronation Street, Boris Johnson, Bakewell tarts and chip butties. (It isn't).

I'm not so far away as the African-bound swallow flies, just 1000 miles south of the white cliffs, just beyond the peaks of the Pyrenees; nothing like shipping out to Australia or Canada. But, it is a sizeable jump in a host of ways under the general heading of cataclysmically changing the course of your lives, and those of your children.

Are you curious as to why people should want to - or are you among the countless who, understandably, wonder if fulfilment lies somewhere far away, while having the good sense to realize that how it comes across on the television can't be for real? Or are you among the hundreds of thousands of ex-expats who have returned home having given it, whatever "it" is, a whirl and found it wanting for emotional, social or financial reasons?

A blunt word of advice if you are determined to fly the coop for a while, or forever: Gird your wotsits. Don't burn any boats with loud boasts and Britain-bashing. Pack an iron-will, a large helping of patience, the appropriate dictionary and, most crucial of all, your sense of humour. Because the fun is about to start.

mother's garden

TICK, TOCK

*Time passes unhindered. When we make
mistakes, we cannot turn the clock back and
try again. All we can do is use the present
well.*

Dalai Lama Tenzin Gyatso

I dipped my hand into the mirror water of our spring-water reservoir one shortening day, and the allure still lingered. It was September 16, 2009. The children had just returned to Latin school. Summer may officially have fizzled out and the dawn air had the zest of autumn, but stone walls, water and brows continued to soak up the pulsing heat.

Daybreak had brimmed with the whistles of the bee-eaters and the warbles of the golden orioles who would soon be gone to Africa. To the east, the crest of firs on the limestone ridge flamed with first light, while dew-darkened vine leaves were fringed with the first red of the impending fall. The almonds were gathered, and any day we would be in the happy muddle of an early grape harvest, such had been the relentless season of ripening. No griping. The grapes were free of the problems moisture brings, and we had cleared the barn and washed the barrels in readiness. We were unbelievably organised, spurred on by willing helpers, and I kept drifting into the barn to savour the view of the long-lost floor. Yet, we asked ourselves again, how long can we keep this up?

Shaking The Tree, Mother's Garden – The Growing Years; honing the art of growing the fruits of oil and wine, watching our children grow; growing a business, growing older, perhaps wiser, and certainly less able to balance our dreams for this little organic farm with the physical challenge. This is the story from 2003, from the last page of *No Going Back – Journey to Mother's Garden*, to 2010.

The children's three-month holiday ended on September 15 and we attempted to regiment some semblance of urgency. Fourteen year old Ella said 2009 had been her best summer "ever". Nine-year-old Joe Joe, a babe in arms when we came, concurred with a dimply grin. This is on account, no doubt, of the comings and goings, the incessant balsa ducking and diving and the harmony of bare-footed young people losing track of time at the piano or with guitars around the garden table. The echoes reverberated, and Ella was drawn more than ever to make music, whiling away great lumps of time at our much preferred (non-computer) keyboard, learning and belting out her own compositions mingled with the Jonas Brothers' greatest hits.

What would you prefer to swim in - a vast circular, nine feet deep reservoir (called a balsa) replete with goldfish, frogs, and the mystery of what else, or a crystal clear, mildly chlorinated, extremely boring rectangular swimming pool with shallow end? The surface of the balsa was rarely still for weeks, and the valley reverberated with shrieks of unbridled glee. A month on from our summer splash and five yards from the kitchen window the purple tent slowly deflated as September breezes unpicked the bent and feeble pegs that could barely engage with the iron earth. With one end tied to a plum tree for security, this was teenagers James' and Stuart's lodging. I was reluctant to take it down, just as was the erstwhile owner who abandoned it at the Glastonbury festival a few years ago. It was a reminder of a Mother's Garden medley of musical, dreamy days when we numbered twenty one and everyone pocketed treasures of time, and orderly queues formed outside the farmhouse's only loo.

I'd like to think that the teenagers may have found here one of those gilded pieces of the jigsaw of identity, for that is what formative experience profoundly is. We all had them - didn't we? – happenings that encompassed all manner of contentment, emotions and learning; seconds or greater slices of our history when we gathered experience and pace en route to adulthood.

Stuart and James came because I worked for many moons on the Eastern Daily Press nightshift with James' father Simon, someone of like-mind, a friend. For all the jaundice and nausea the celebrity-obsessed media may increasingly engender, I remain fiercely proud of that newspaper and am forever appreciative of the ethics, keen minds, dedication, and friendships of the people who saved my face more than once when I was supposedly putting the paper to bed. The same is true of the Yorkshire Post, an equally important English morning newspaper of identity, for which I have written for six years.

So, into the pantomime of Mother's Garden these two young friends were plunged, expecting (they admitted too late) more hard labour and less diversion. They gathered almonds, attacked brambles, raked fallen figs, washed wine bottles and waited willingly for more instruction as we schooled them in the art of distraction. One balmy, back end of a day of doing precious little, up stepped Jeannie from north London, the twelve-year-old singer-songwriter, who radiated back the dreamy magic of the coloured lights that dangled above our heads between house and trees. We all fell silent, intoxicated by prodigious talent and farm red wine. In the mix were Jeannie's guitarist brother Callum, her musical mum Cathy and dad Neil, Harry our summer/autumn farm helper/guitarist, his sister Grace and her man Ross, another family from south London, some of our neighbours and my young cousins James and Becky. James spent a while with us, having stepped away from chasing life, career and his tail in England. In his time here he profoundly helped in all manner of ways, and there was a great void when he left embarking on walkabout and a journey of personal discovery in New Zealand and Australia.

Away on the hill the village fiesta had come to a cringing conclusion. Joe Joe's marching drum band had been rehearsing for days with a rock ensemble who were billed as the cherry on the top of a week of frivolity. Having sat through several run-throughs of unthinkably horrendous noise I knew what was coming, and warned anyone who would listen. But, the rock band had a cunning plan. With thirty rhythmic villagers on stage a throng was guaranteed.

The moment Joe Joe and the drummers left the stage, however, prompting the rockers to let rip with a racket which sounded like four heavy metal solos being played at the same time but on different planets, if you get my drift, anyone able to walk legged it.

In the nearby town, the firemen went home happy after the usual explosive end to their festivities passed off without major incident, despite the alarming spectacle of a dragon with fireworks in its mouth spouting sparks into the air and through open windows in the narrow streets.

So endeth summer, riotously colourful with the splash of youth.

All that was needed thereafter was much improved rations of sleep. What we didn't need was a double helping of trouble in the middle of the night. It kicked off just after one o'clock in the morning, when something suitably grim sent the dogs into delirium. I slipped out of the back door and stood in the moon shadow cast by the wood store. A grey lump pranced through the night, weaving between the hazels and straight through our vegetable patch. It was either an enormous albino boar or a horse. *Merda*. Worse – it was a colossal, spooked stallion with a Marty Feldman Young Frankenstein expression. Two more circuits through our tomatoes, cucumbers, courgettes, peppers and aubergines and we'd lose the bloody lot.

Pens and corrals of varying degrees of effectiveness can be found across the valley and there have been the occasional nocturnal breakouts and visitations. Nothing as troublesome as the boar. But, I hadn't seen this equine brute before, and unlike the others it clearly wasn't just galloping through en route to the horizon. I employed the usual tactic of flapping and clapping to imply discontent, but the gangly maniac merely charged round the olive grove and remerged out of the blackness with a fiery little speckled pony which proved to have an even worse disposition.

The pair danced across the vegetables again for good measure, then belted towards our corral and our pony mares, La Petita and Remoli, where, in a cocktail of lust, rivalry and anger, the visitors proceeded to kick the hell out of each other and then smash their way in. I chased the horse out, but the vicious pocket-rocket pony stallion wasn't having any of it. It wanted satisfaction and there ensued a three-hour on-off melee of kicking, biting, bonking and screaming while I endeavoured to stop our two abused females from belting out through the gap. Maybe I should have let them run to spare them. La Petita was standing her ground and doing her best to protect Remoli, her daughter, yet the longer it went on the more we feared they would be injured. Long ago, though, I'd encountered the shocking carnage where loose horses and cars had collided. I couldn't take the risk of them galloping on to the road.

We rang neighbours. I steamed over the hill to where the shepherd lives and keeps an assortment of horses, but all to no avail. Nobody had a clue who owned them.

The Catalan police were called in case anyone had reported the loss. Three freshly ironed young officers came at 3.30am and obligingly joined in the fray despite their obvious lack of training in equestrian psychiatry. They helped to keep the interlopers at bay while we harnessed and led our mares away, then the officers began making calls of their own, again without success. And while all that was going on a boar, oblivious to the mayhem, bulldozed its way through the undergrowth on the terrace below us. Our ponies were now out of the firing line, but we had nowhere to put them save in the barn, with the barrels and cases of 2008 vintage wine we had bottled that morning. Within thirty minutes the scent of grapes was a distant memory.

What a night. By 10.00am the owner had been traced and life settled again into some semblance of bleary-eyed calm, for a few days at least. The foal consequences may have already been established, but just to be sure the pony stallion returned sixteen times over the next five months until, inspired partly by our anger and finally, my kneeling and begging, the weary owner finally accepted he'd failed spectacularly and removed Rodeo Romeo from the valley. The exasperating saga had included me standing guard from 11.00pm to daybreak one very long, cold January night, and another deep chill experience when, in knee-high snow, I'd managed to collar the brute only to be dragged horizontally until my jacket and wellies were jammed full of the white stuff. By the following Easter it was obvious that both ponies were expecting.

2009/10 was to prove to be one of the hardest of winters in many ways.

We lit our first fires with the earliest ground frost to date, in mid October, which whacked the courgette plants, had us fishing in the winter trunk of scarves, gloves and hats, and made me anxious about our woodpile. Harry had been sawing furiously all summer while putting his RSPB training to use, clearing pine woodland to create a more attractive bird habitat. Log stacks abounded. But you need to feed our wood burners like a stoker on a steam train and the early call for heat boded accurately that we wouldn't make it to March without buying in some seasoned chunks of iron-hard almond. Lightning flashed, the trip switch at the top of the land capitulated, and to add to the cold we had our first tempest of the season.

A week later and - oomph - we were back to thirty degrees. Nell the tractor was on the blink, so I borrowed Mac and Conxita's old Fordson to plough the olive grove in readiness for the winter veg. The two farms lie about two miles apart as the raven glides, a bit farther when you have weaved along the old tracks between vineyards, across abandoned almond groves and past orchards and banks of wild cane.

The ploughing done, Joe Joe drove the Fordson home, with me and a sorely needed cushion perched on the side, close to the clutch and stop button and within reach of the steering wheel which has a habit of coming loose. We stopped to scavenge boughs of dead almond as a thank you for the loan, then with Mac's guidance carefully cleaned and oiled the blades of the plough before it was laid up for winter. Cake and kisses, and we set off for home, waving away Mac's offer to run us back in his Land Rover, me with my cushion stuffed down the back of my grubby overalls like a geisha mechanic.

Then the winter sank her teeth like never before.

There is regular talk of this Mother's Garden life being a kind of time travel. I read the likes of Lilias Rider Haggard and Adrian Bell incessantly, and their early twentieth century accounts of a distant rural England and customs hold a great deal of what we seek, not least pace. I love to walk between the farms, alone or with my family and friends, and I never tire of it. Mac has just read my copy of Bell's *Corduroy*. I am re-reading *Men and The Fields*, written in 1939, where, among the watercolours of words, Bell gently questions.

"How little people walk nowadays, even in the country. When you say, 'I'm going to...' naming the next village, two miles away, it is taken for granted that you are going by car. The surprise when you set out with a stick on foot is itself surprising. 'What – you are going to *walk*!' It is not only that you should have the inclination, but that you should have the time. To walk two miles there and two miles back is equivalent of wasting the afternoon. The modern farmer is right: he cannot afford to go there on foot, in the same way that although to drive a horse and trap is infinitely cheaper for him than running a car (because he grows the food for his horse but has to pay a retail price for petrol), yet it is dearer for him in the long run, because all other farmers, his competitors, run cars, and he must work at their pace or go behind So one sets out on foot, quite aware of the indulgence it is."

What is Mother's Garden? A delicious muddle, someone called it. That's our life in vivid colour, where all comes to pass, somehow, sometime. For me Mother's Garden is a journey without moving. There is no telling from one day to the next where we will be carried, what we will face and whether we can cope physically and emotionally. There is a school of thought that, for all the talk of undeniable feasts for the senses, we have bitten off more than we can chew. I have challenged that, but given the ageing process and, incompatibly, our fallible inclination to weave an ever more complex existence, we have begun to talk seriously of surrendering something, whether that be land or labour or both.

Ten intoxicating acres run from rough pasture by the lane to ragged and dying almond grove at the crest of our modest hill among the mountains. Peregrines look down on lines of vines, an avenue of olives smudged with a wilderness of hazel crop abandoned to pine, holm oak and dog rose, all overlaid with the music of spring water, barking dogs, puffing tractor, and infinitesimal birdsong. I have built three crude benches and have a mind to make a dozen more. One sits three steps from the kitchen window, beside the graves of our springer spaniels, Charlie and Megan, who came with us from England in January 2001. It's shaded by plum trees, two colossal Cyprus sentinels and the twisting wisteria that clambers over the back of the house. We cup mugs and look north, across our variable vegetable garden wedged among young olives, to the walnuts that shade the old caravan, the dwelling of the young volunteers who have journeyed here off and on from America, Canada and northern Europe. The two ponies graze tantalizingly close to the tomatoes, keeping the growth beneath their hooves at bay, while above their backs the olives cloud with happiness at the poo they leave behind. The second bench is by the veggie patch beside the spring, and the third is way up the track, on the brow between our old *carinyena* vineyard and dying almond trees, where I am convinced an ancient dwelling once stood. Feet constantly trip on worn rocks with right angles and perfect corners. I carry them around while scanning for more, then sit and slide my gaze from the Stone Age cavern in the red rock across the valley to the village and coned peak beside it where the Romans built a temple and villas. With a spring on our land, and the easy rise of it to sandy crown, imaginings of earlier dwellers come freely, and I feel it. If only the earth could talk.

East of the old house sits our pretty square of three hundred and fifty *garnatxa* vines which in our first heady season we weeded by hand. It makes me chuckle just to think of the nonsensical satisfaction. To the south and front, beyond billowing figs, walnut and medlar, the rhythm of persimmon, pear, quince and apple trees is broken by our chicken coop. The hounds have the run of the farm in the morning and evening, but the afternoon belongs to the hens, and they fan out into the orchard or spin round the back to check out the compost. To the west, beyond the barn, our holiday cottage looks to the village and the setting sun, where anything from two to ten people have dwelled awhile. The flourish of faces is all part of the wild swing of the summer-winter pendulum, when we go from crush, swelter, fiesta and siesta, to solitude, double duvets, damp firewood and deliciously clear and crunchy dawns; from forty degrees to minus fifteen.

In some ways it's not so important where it is, for this life, this book, is as much about attitude as latitude. But in mapping terms it's a ten acre patch, one mountain range away from the Mediterranean, in a verdant Catalonian valley where Romans planted vines and olives and where some aspects of land and values have little changed in 2000 years. Listen and you will hear birdsong and the banter of chickens, children and puppies, the bells of a neighbour's goatherd, jam bubbling, figs falling, the spring gurgling, me bumbling, bee-eaters whistling happily over our hives, and the pines whispering in the breeze. Look and you will see either the fine art of being or a fine mess, depending on your state of mind. The eagles and ravens gaze down on our chickens and a red-soil mix of care and wilderness, of vines and potato patches crisscrossed by the trails of wild boar, the huge round mirror of our reservoir rippled by drinking swallows, the broad backs and broader tummies of two grazing brown and white ponies, an arthritic MF tractor with tyres of varying baldness, untidy coils of irrigation hose locked by fennel, and a mustard yellow football blotching the rhythm of the farmhouse's terracotta roof. A moss-covered Range Rover, the wagon of emigration a decade ago, now the abode of Jess the pink cat and our rat-proof store for fowl feed, continues to wait in the shade of one of the fig trees for us to scrap it. Assorted building materials squirreled from the dump are piled artistically against the dry-stone barn wall, while under the walnut tree-house stands a pretty kitchen chair with a broken wicker bottom. That just about sums us up.

Venture a little, along the lane and into the shallows of our village, or knee-deep into this mountain backwater called the Priorat, and you may begin to appreciate another reason why, for all the self-doubts and questioning, we are still here. There is something about the glue of Latin life. It has held through the millennia and bonds with the power of family, contact with the earth, respect for the sun and wisdom of eating sensibly.

Sensibly? Provenance, sharing and time. We talk of another book, the illustrated Mother's Garden *Survival Guide*, a literary feast of recipes, the first of which is inedible. I'm talking about tables. How significant is the breakfast/lunch/dinner table in your home? Do you share meals daily? We have agreed that 'recipe one' will be the essential ingredient of all meals that follow, namely the rearranging of furniture and priorities, and placing the family meal table at the heart of everyday life, out of vision and earshot of a television. We have two tables, my grandparents' mock Tudor oak refectory eight-seater which dominates our kitchen, and a garden version in the dappled shade of a fig, the latest in a long line of free furniture from the dump.

We so enjoy cooking on the red soil over fires of dead vines – we always lose a few plants every winter and the smoke imbues such delicate flavour – and then sharing food with good company in adherence to the maxim that what sunshine is to flowers, smiles are to humanity (Joseph Addison). 2009 presented a problem though. The two smaller, knackered old wooden tables which were relegated some years ago to outside use, finally keeled over after first one and then the other lost a leg. I propped them up against each other for a couple of dangerously unstable al fresco feasts, then resorted to a plywood-between-trestles contraption, but that drooped when loaded and lacked soul.

It can easily be argued that everywhere you turn there are more important things to be doing than crude carpentry that swallows the best part of a day, but we needed something with character and four legs to bring colour and promise. Anyway, it was fun creating a garden table from junk. The worm-riddled base had been fished out of a skip, its turned legs catching the eye during a back-street hunt for a parking space in the city of Reus. How pretty we looked with that strapped to the car roof. How pleasing to the eye it is now, treated and painted sky blue. The table top was made from a pile of fencing planks. All good things come to he who wishes, and I tugged them from the rubble at the tip two weeks after I began scavenging in earnest. The top has the faintly discernable curve of the earth, plus one leg will forever need a brick under it (which reminds me, I must cap the feet with old sweet corn tins), but set on the rough grass in the lee of the house and ringed by an assortment of similarly rescued cane, metal and wooden chairs it is so pretty, and might feed us and conversation for a few years.

It isn't that we have more time (anyone who is self-employed knows the score). It's just that what consideration we give to food has gone to the top of our priorities, from the joy of growing things, to the importance of savouring the conclusion. We work hard to make this happen. For the rural people here across southern Europe and probably in all communities closer to the earth from Siberia to Patagonia, it's ridiculous to think otherwise. So I wince every time I pass the petrol station at Reus where there is an *On The Run* cafe and shop. Most Catalans won't know what it means, but I do and it riles me. It is just one snappy commercial example of a clever world forever finding ways to save time. It fits too, among the fast food outlets and spreading urban Spanish dalliance with the hurrying habits of the juggernaut economies. Maybe such convenience is very necessary. Any break is better than none, and it may deter people from attempting all manner of time-saving things while driving. "Hello", incidentally, to the woman I used to follow in to work occasionally when I was a UK commuter, and who managed to do her hair and make-up while at the wheel. "Hello" to the man who drove while sipping a steaming mug of tea or coffee.

I understand economics. Money sort of makes the world go round. The faster we can make it spin, the more we can have. Do you stop to eat, or are you getting more things done by working through and snatching something on the hoof? I did it, cursing when I had to waste a minute at the pedestrian crossing on my fast walk back to my desk from the sandwich and chocolate bar shop. Computer keyboard cleaning companies must be making a fortune. And, hand on heart, if I was plucked from here and plopped back in midstream, I'd be at risk of re-offending my health and well-being again, unable to unplug my anxiety at getting the job done because, well, the compulsion and expectation is to press on.

But it is not just in the workplace.

It is madness and, I firmly believe, the root of many issues, how the chop-chop-busy-busy, tray-in-front-of-the-TV, oh-so-advanced first nations have chosen to devalue the simple truth, that good food shared fosters wellbeing, happiness, openness and harmony. Think about it. There is no better time to talk through difficulties, underpin family, bridge generations and practise the art of expression and listening, than when tucking in. It's when basic manners and courtesies are fostered, when bonds are reinforced.

We are putting it to the test. We are a family facing the universal issues of adolescence. Ella, five when we veered somewhat recklessly off the beaten track, is taller than her mum now.

Fifteen is the age of wondering, isn't it? Those first serious ponderings beyond the range of vision to the peripheries of heart and soul; of self-consciousness to the point that walking into a room of unfamiliar company is the greatest of challenges; of mirrors and imaginings; the uncertain destiny of a fledgling edging along the branch.

Quite how to handle this as a parent is an eternal question. We can, perhaps, recall from our dim and distant histories the teenage muddle of emotions and erudite attitudes. Watching your children grapple with this you have to play the grown-up (even if there isn't such a thing), shape your sentences, try not to take all things personally, and to accept your fading light. Perhaps not fading, just diminished in the growing company of flares and flames that illuminate a young life. The hope is you have done enough to make yours the constant, certain light. Someone said just the other day, when Ella was but our babe in arms, that if you have not instilled core values, love and trust in offspring by the time they are rising to teens then you never will. We are at the point of finding out.

Here lies one of the largest planks in our reasoning to come. I have been permanent, if not always constant. Maggie and I parent in partnership. Since we shook up our lives I have missed nothing of Ella and Joe Joe's childhood, and the poor things have missed nothing of my middle age and what I truly am (repetitive, disorganised, mathematically incompetent, occasionally grumpy, and prone to spend an inordinately long time on the loo with a book). Foibles are bared, not least in the linguistic sense, so remember any of you heading in this general direction with offspring, that you are about to face something few parents have to – daily, unavoidable inferiority. Ella and Joe Joe's world, as is ours, is dominated by the indubitable truth that Mum and Dad will never, ever, despite their best endeavours, be effortlessly fluent like they are (Spanish and Catalan). We need their help occasionally, while they have to live with the shame of us umming and erring, or, worse, me making a baboon of myself on local radio and television. All of which raises unavoidable anxiety and debate about identity.

Through our unusual life choices we have presented our children with something else to get their heads around. They are not sore thumb foreigners (incomers abound and their schools have children from across Europe, North Africa and South America) and they openly declare and show they are more than happy with their lot, thank goodness. But they have no roots here. It isn't easy being different, and no child would choose to be, so when they have very occasional bad days for whatever reason and our already heightened anxiety goes into orbit at the slightest hint that they may be suffering a sense of alienation, we lay in bed wondering how hard we may have made it for them to get their bearings in later life. Normality is, I suppose, having the security of the familiar and a family base where you can "go home", whatever your age. But there is no saying where Maggie and I will be in our dotage. We talk of setting sail around the Med while we are still physically able; of Indian ashrams, of the colours and scents of every continent on this earth that await us. We feel sure that "home" will be wherever we are, and that might just be Mother's Garden, but with nutty parents like us, the extended family and awareness of who we are and where we come from, are all the more vital for Ella and Joe Joe.

So good it was, then, in the summer of 2009, to wave Maggie and Ella off at Barcelona Airport, knowing they were going to England to share some time away from routine, while looking into the past and possibly the future.

First London. Mother has gifted daughter her love of the arts and there were many stories to tell and places to see. Thanks to vital musical friendships there was the rare treat of a Royal Opera House backstage tour and best seats for a dress rehearsal of *The Barber of Seville*. Then, while riding on red buses and photographing telephone boxes, Maggie showed Ella her old college, Froebel at Roehampton, and the Wigmore Hall where she worked when she was twentysomething. The Wigmore had been a huge stepping stone, the flowering of Maggie's knowledge and love of chamber music and which would lead to other important ventures, and to me.

Family next, and they journeyed to the northern cusp of the capital, meeting up with Maggie's sister Liz and mum Beryl in Enfield, to glimpse the green belt farms – Park Farm, Ferny Hill, Barnet, and the neighbour Parkside Farm, Hadley Road, Enfield – which Maggie's late father David Whitman and grandfather once ran, and where she, like Ella, had the indelible goodness of an early childhood close to nature.

Further north in leafy Hertfordshire, at Great and Little Gaddesden, Nana Beryl in turn showed where she roamed the pastures and woodland as a young girl. They picnicked in Ashridge Park where Norfolk-born Godfrey Bunn, Beryl's grandfather, was groom and later coachman. They stared at old family houses, visited Whitman graves (coincidentally on the birthday of another great-grandfather, George Whitman) and talked of the voices long gone. It was an important start. But lineage on both sides of the family is a colourful weave, and we have yet to fully explain the roots that run from Suffolk, Yorkshire, Devon, Shropshire, Leicestershire Scotland and Wales. It is said, too, with certainty but no detail, that both our families have an ounce of Spanish blood. Maggie, born in Canada, was raised in Hertfordshire and Norfolk. I consider myself a north Norfolk boy, raised near Holt, christened at Glandford and grafted through birth, love and experiences to the swathe of land, sand, marsh, sea and sky between West Runton and The Wash. But that is just a happy coincidence of fortune, as I hope Ella and Joe Joe will feel about being here in Catalonia during their formative years. We are commonly of the one planet after all.

"I am English," Ella said with consideration on her return to Mother's Garden. "But this is my home."

While they were away, Joe Joe and I enjoyed some manly one-to-one and somehow muddled through the chores. We taught our two terrier puppies – Tilly and Ted – to sit on command. We watered to the point of vegetables, chrysanthemums and herbs drowning rather than dying of thirst. And we tended to our holiday cottage guests, picked fruits, cleaned the corral and managed to get the house back to some semblance of normality on the cusp of the women's return. We even managed a musical diversion of our own.

Sitges fizzes. It's a large seaside town of some beauty, just south of Barcelona, with a warren of narrow streets and the chemistry of contrast. There is always something happening, and on July 11 it was an evening drum festival featuring, among six hundred others, Joe Joe and his fellow village *tambors*.

Through the streets they processed, rattling dentures and dislodging toupees – hip-swinging samba groups and our distinctly less animated, but no less rhythmic thirty-strong village band with almost as many proud family members swirling about.

They thundered away for nearly two hours, from the seafront crush with its colourful characters (man in leather, studded dog collar and goat on lead) to under the railway arch where everything amplified to fingers-in-ears intensity. There was just one log jam, outside a shop called Love, Sex and

Diamonds, when a few of our entourage were lured in, which was no bad thing because it gave one grandmother in our party more time to press on with her labours. As we marched she busied herself with wonderful country care, draining bottles of water left for thirsty drummers into the dry flower tubs that decorated the route.

Shaking The Tree; Mother's Garden – The Growing Years takes up the story where *No Going Back – Journey to Mother's Garden* left off in 2003. I use the term loosely because, of course, I navigate through events and thoughts in my usual haphazard way, while, hopefully, giving an honest sense of the rhythm, humour, anxieties and wonders of existence.

Mother's Garden? It is, for sure, real. It's family, people, tears and laughter, blisters and the bitter cold of January doubts. It is the fulfilment of a belief that life can only be the sum of experience, and nothing is gained without venture, living in the moment and following dreams. It is parochial, impossible, unforgiving and utterly enchanting.

THIS WON'T HURT

Consider, gentlemen, the fate of any
Englishmen caught by the rebels in the pass
of Killiecrankie.
   History teacher "Tweedie" Harris (circa 1975)

Let's skip the niceties. This book is painfully honest.

It has troubled me to think what my mother-in-law, readers of my relatively pastoral chronicles in the Eastern Daily Press and Yorkshire Post, anyone for that matter with glorious English reserve, will think of what I'm about to recount. So, if you don't want to know the result, please look away now, or spin forward to chapter three.

I toyed with the idea of calling this chapter LIFE'S A RIALLETA, Catalan for laugh. But when you are wearing nothing more than a bare-backed surgical gown and hairnet and sitting pink-cheeked on an incontinence pad while awaiting your turn in a foreign vasectomy queue it most definitely isn't. (I did warn you).

Rule number one for all high-tailers. Never try and wing it linguistically.

I had a shrewd idea what I was in for, of course, but my reckless tendency to nod sagely even if I couldn't be absolutely sure what was being said meant that I'd left the Spanish surgeon's office two weeks earlier with precious little understanding of what he and a legion of young nurses were going to do to me. Maybe that was not such a bad thing after all.

The whole neutering episode had, up until the incontinence pad moment, been one of those dangerously hazy chapters when, in a fog of ignorance and resignation, you put your faith in the gods and just hope Hades isn't on duty. All straight forward, they said. Nothing to worry about, they said. I managed to grasp that much. And given Spain's pretty efficient health service a Latin vasectomy had seemed (after several stiff drinks) a fairly reasonable suggestion by my wife Maggie Whitman, the mother of our gorgeous-but-two's-enough pickles.

Mind you, so did putting on my deeply scratched and virtually useless ski-goggles and whizzing the strimmer over the patch of tall grass outside the kitchen window, shortly after Petita the pony had deposited some of her squidgy heavy-duty fertilizer. I was blind to the consequences, and had been lulled by the gentle smile and restful music being piped into the consultant's office as he wasted his time trying to describe my re-plumbing.

So, all but naked, there I sat for nearly an hour, waiting my turn under the knife while Maggie waited downstairs, leafing through a dog-eared copy of Hola! magazine like a mother in a hairdresser's. I stared at the wall, with three young Catalan nurses at their desk out of the corner of my left eye, and three other hairnet suckers out of the corner of my right, trying like a boy on a rock-hard church pew not to move a muscle. If I did, the pad would scrunch and squeal in protest and all heads would turn. I did risk checking, though, that it wasn't just me who had been placed on a large square of absorbent material in that final hour.

When I'd edged into the waiting room, back to the ice-cold tiled wall, I'd tried a manful wave to the other guys already there, but it fell way short. A bearded bloke I remembered seeing in the consultant's waiting room sort of winced a reply, while the condemned next to him continued to stare mournfully between his knees. We were, but for the remarkably cocky and extraordinary youthful lad at the end of the line, who was tapping the tune from his MP3 player out on his thighs, like naughty schoolboys waiting to be flogged.

What's the worst that can happen, I consoled myself. Nobody here knows you. You are in a foreign country for goodness sake, so absolutely no chance of a friend-of-a-friend's fate who, while lying stark 'b' naked on the operating theatre, suddenly realised the nurse beaming down at him with a twinkle in her eye was the woman who lived four doors along his street. The thought suddenly made me feel a whole lot better. I glanced at the other guys and found myself trying to lock in a giggle like you do a sneeze or wind, in one of those completely inexplicable irreverent moments, like laughing at a hot-air balloon in the shape of Bernard Manning even though it isn't funny.

I yoga breathed, tried to watch the Spanish television soap opera with wobbly-scenery on the screen above my head, thumbed through a couple of magazines to take my mind off the itch on my left buttock, and gained further composure by reminding myself what friend Johnny B had told me.

He was the only close friend who had had the snip.

"Oh, nothing to it really," he'd said stiff-upper-lipperly. "A mild irritation that's all."

"Really?"

"Absolutely. Couple of days and you will be fine." He'd then coughed in an enough-of-all-that English sort of way and changed the subject.

When my turn came I left the waiting room far more brashly than I'd arrived, and didn't care anymore who saw my bottom. It would all be over in a jiffy and I'd be home within two hours.

It was a quarter to one. I was the last to be rewired before lunch.

I bared all, laid back and tried to think of England. We'd be back there in a few weeks, sailing on the Norfolk Broads and - OOMPH, in went the local anaesthetic, only to be expected, stay focused – and we could head for our usual peaceful haunts in the reeds and spend a day reading books.

I concentrated on a few more favourite things and the surgeon whizzed along the side of the operating table on his wheely stool to smile at me through his mask with a nod and a wave of the scalpel to suggest he was ready.

Fair enough, I thought. No going back NOOOOOOOOOOOW! I hit him with several bars of the Hallelujah Chorus. He whizzed back. "More anaesthetic I thinks," he said with laughter in his eyes. I could barely breathe.

Back home I lay on the bed, legs akimbo and asked Maggie to fetch me a brandy and the telephone. Johnny B's wife answered. The topic was too acute for me to be ambiguous.

"Did he say that?" She asked.

"Yes he bloody well did."

"Blimey. You should have seen him. He was waddling about bow-legged for days."

Talking of things that make me wince, I need to talk to Jeremy Clarkson.

We went on a day trip to Italy together for lunch once, as guests of Fiat, not that I expect him to remember. It was when I was a motoring writer back in the daft eighties. At least I think it was him. That was when car companies went to gloriously lavish extremes to impress an unlikely jumble of British journos who had somehow convinced their editors that they should be given the keys to a new supercar every week and bunk off work to eat caviar while watching the sunset behind the Pont de Vecio in Florence. Oh, and write something about smooth gearboxes and under-steer to fill the spaces between the lucrative advertising in the motoring pages.

You had to be a tad cocky to get the job, but if you did, the rewards were, well, ridiculous.

I say unlikely bunch, because when I went on my first jolly, er, I mean assignment, there was still a couple of grey-gilled chaps in cravats and driving gloves, but the wise-cracking young petrol-heads were taking over.

We buzzed about the globe in private jets, with big hair, button down collars, spray on jeans and bulbous opinions. No change there then, Jeremy. We whizzed along Highway One in California, did handbrake turns on Finnish rally stages, were given racetrack training by the incredible Jackie Stewart, played the casino in Monte Carlo, and even went all the way from Gatwick to Italy for lunch to see the spotless factory that was to build the new Fiat Uno. Yes, it was that long ago.

If you dented something they gave you another one. If you needed a helicopter link from Heathrow to Gatwick, no problem.

It was the perfect training ground for anyone with the wit and nerve to make a career out of driving unthinkably expensive cars into swimming pools and being so successfully and (many feel) refeshingly un-pc. The Clarkson column in the Sunday Times can make me spit feathers, but every time he takes a seed of truth and grows a triffid, and I raise my hand to slap it over my open mouth, I remember my red-blooded, adrenalin fizz, laddish, long-forsaken motoring background. I never took that too seriously either. He can be hilarious, of course, which compensates most of the time for ludicrous sweeping remarks. Like the quip that ex-pats all wear loud shirts and prop the bar up from 10am. Excuse me.

For sure he won't recall our chat at Pescara airport, but he might remember Fiat's outdoor buffet in a howling gale and the pilot of the old BAC 111 jet declaring he wasn't sure the runway was long enough for us to take off again. Either way, I'd like to offer Señor Clarkson a lift from Reus Airport to Mother's Garden in our Opel Zafira, bought on his recommendation, to point out the obligatory Seat 600 rusting in a corner of every farm, and put him straight on a few expat points. He'd ask, no doubt, why I jacked in the company car, the pretty cottage and the country to take on an overgrown farm and live in a ramshackled and draughty house in a country where they talk funny.

Tricky.

The British media has been digging furiously for an answer to the great exodus from the UK, beyond the obvious answer that the globe has been in massive unavoidable transition, both climatically and culturally, and now financially, of course. Unsettling isn't it? People have been sucked back from the Costas to Suburbia on account of the dire exchange rate, but I bet your bottom Euro that the moment things flip again the ferry ports will be full of daydream believers.

In 2007, I flew back to see my dad and to peddle some scrumptious fresh olive oil, and then Maggie joined me for the last couple of days so we could attend a wedding. As I waited in the arrivals throng at Stansted Airport I realised how radically British society had metamorphosed since we bought our Catalan cabbage patch in 2000.

I stood among the crowd at arrivals waiting for Maggie to come through. The plane was late, so as I kicked my heels I sifted through the potpourri of voices around me. The couple to my left were Portuguese, I'm pretty sure, because the language is very similar to Catalan and I could half earwig that she couldn't wait to see her parents; beyond them two east European men were talking incessantly and aggressively on their mobiles; and to my right and behind me stood a great gaggle of Poles clutching bunches of flowers and some flags. They were mostly young men, a couple of women and a sprinkling of children.

Then she was there. Not Maggie, but the vast Polish grandmother, waddling, overwhelmed, into the arms of her family. Flowers and tears were falling in all directions. The party was even bigger than I'd first thought and other family members jogged from the coffee shop to join the giant hug.

Airports here in Spain have long been packed with immigrants talking funny and with money in their pockets – tanned, early-retirement Brits, Dutch, German, and Belgians in ironed shorts, white socks and sandals. But I was stunned by that Stansted experience. I shouldn't have been, because with a larger and more diverse European Union people were bound to flock like moths to the brightest economic lights, as would we all, of course (come on, admit it), were we in their shoes. Maybe they are back home in Poland now, given the economic resilience there and the squeeze and depression in England.

What fazed me, though, and still does, is the pressure of such rapid change both on society and, in crowded, wonderfully liberal and welcoming Britain's case, on fundamental services. Someone has the measure of it all, one can but trust, and social history tells us that the populations have always ebbed and flowed, drawn by economical and political tides. But every time I have returned there has been a palpable tension, a diminishing openness yet a readiness to talk of disquiet. This may have been amplified by my relative isolation, and be a muddled consequence of many things like afluenza (look it up), hamster wheel exhaustion, or the endless tirade of wholly depressing news, but the drastic immigration shift couldn't have helped.

As just one ant wandering upon the round rock I try to remember when I see people striving for a better life that there, too, go I. And when my Catalan neighbours look at me, an immigrant, I hope there is a willingness to see an individual not a stereotype. Someone brought me a newspaper a couple of years ago that saw fit to contrast on the front page a large photograph of what looked like a group of shifty unwashed eastern European men loitering on a street corner with another image, equally big, of a manicured English family sitting on a manicured lawn. Influx and exodus was the paper's angle. The story was built on truth but the choice of images did nothing but feed fear and prejudice. It had, the editor thought, more impact than the other story of the day, the random shooting of a boy in Liverpool. The Poles I spent an hour with at Stansted Airport, whatever their story, seemed a loving, law-abiding and caring family. There are, no doubt, British ex-pats who are not.

I grew up in a cocoon of relative comfort, complete with cheese plant in the corner of the living room while Kojak sucked his lollipop on a new-fangled colour television; little me, watching another world and all its woes from a vast armchair, then downing a glass of milk and a home-made biscuit before going out to play in my real world, a seaside town at the north end of the cul-de-sac called Norfolk. Yes, Jeremy I know what you think of Norfolk.

It is a fair argument that I have back-tracked into another bubble, for there is something of a sense of stop-the-clock about our lives. We often draw comparisons with what it might have been like in England when all villages still had their souls, shops, crafts and a weave of roots. But Spain is just as much in transition as Britain, with a massive eastern European influx alongside the longer-established north European, Latin American and north African faces.

Another British newspaper argued that the legions of ex-pats, together with the millions who apparently hanker to join them, are dewy-eyed about the 1950s. Really? I was two in 1960, so don't remember. My formative years were the late sixties and seventies, and while I still inexplicably remember the lyrics to Sandie Shaw's Eurovision *Puppet On A String* I'm not inclined to cast my mind back. The heights of fashion in the mid seventies were ludicrously voluminous trousers that looked like Pan's People hotpants with windsocks sown on. I was particularly fond of a dog-tooth pair, which I thought looked fetching with patent leather stack shoes and a skin tight cheesecloth shirt, or (my absolute favourite) a green wafer thin cotton shirt with the largest butterfly collar seen on the planet, all topped out with a mop of hair and facial fuzz that young bucks thought made them look like the bloke out of Joy of Sex. How I ever pulled, I'll never know.

If anything, it was 1930s America that I got nostalgic about, to wit the fictional Waltons, every Sunday evening, to which everyone in the family was addicted, not that I'd have ever admitted it. I can't complain. It was a childhood at the cutting edge of modernity, posh some might say, in that my mother's Honeylands guesthouse in Sheringham had an avocado bathroom suite, sinks in all the rooms, and classy cuisine in the shape of curly butter with breakfast, plus poppy seed rolls, pineapple rings on your gammon steak and croutons floating in soup served from a hostess trolley. We are talking liebfraumilch by the lorry load and teasmaids beside every bed, including mine because it was rented out from Easter to late September and I slept on a mattress in the dining room.

Mum was committed to giving the impression we were more than middle class, although for the first couple of winters after she had left my Dad in 1965 and set up home with Horace, we had to collect cones in nearby pine woods to keep the Rayburn alight, and we always free-wheeled the old Anglia down hills. Yes, there are hills in Norfolk. It was four years after that, when Honeylands was packing them in, and my beaming mother suggested (just the once) that we all dress up as bees in the carnival for a publicity stunt, I got my first taste of Spain. After Mum and Horace's third lucrative summer season giving people from Leicester a holiday to remember in sea-breeze Sheringham, we flew one October with Dan Air to Majorca where my six foot two step-father hired a miniscule Seat 600E, shoe-horned me into the back and then proceeded to tackle the highest and narrowest hairpin bends on the island. I remember the place being relatively quiet and unspoilt, with women wearing a lot less than in draughty Norfolk. The first-floor apartment where we stayed was literally five metres from the shore and Mum would hand sandwiches down from the balcony. Today that old apartment, though still standing, is separated from now distant water by a vast and crowded promenade.

So, what of that newspaper suggestion that people long for the old times? The 1950s theory has possible credence if you airbrush out the grim bits and tweak up a nation's instinctive craving for community, social responsibility and a more sensible rhythm, with more time for the family. Which must make the Latin lifestyle and climate, and doing nothing at all for hours, a tad alluring.

The bottom line, though, is that anything can seem easy with privilege, which in the material world the British middle-class masses have (notwithstanding the work-work whizz bang pace and stress) in abundance. Sell your semi, buy an olive farm and stick a satellite disk on the roof. Swap gridlock on the M4 at Slough or Manchester orbital for waiting for a goatherd to cross the lane to your Mediterranean retreat, without having to give up Top Gear or the Antiques Roadshow.

In our case we were just crazy. It was 2000, before (believe it or not) there were any "New Life Abroad" documentaries. We didn't have enough money, we didn't have any idea what we would do to earn a living, and we had two very young children. What happened, quite simply, was we followed an irresistible urge to veer off the common track. Fate led us to Mother's Garden. Maggie and me like to be spontaneous once in a while - you know, full steam ahead, taking decisions on the spur of moment with little thought of consequences. Like figuring that leaving England at 4am on a bitterly cold January morning in a beaten up Range Rover stacked to the gunnels with children, two dogs, bedding, flasks of coffee (ie moving house and country in the same day) wasn't insane. I've had a few fat slices of second thoughts but, hand on heart, that moment, walking the dogs across a deserted, desolate, colder-than-a-witch's-wotsit Dover Port car park and looking at Maggie and children Ella, five and Joe Joe, six months, huddled in the car, wasn't one of them. We were going for it, come hell or high water, buoyed by dreams, caffeine and incredible faith that all would be well.

Since the television documentaries and the publication of *No Going Back – Journey to Mother's Garden* in 2003 we have received thousands upon thousands of emails and letters from around the world. They are still coming. A few people say their enthusiasm has been fuelled not diminished by my accounts. A great many others, though, say the story, either on TV or in the book, has made them realise that maybe moving away from all they know isn't such a good idea.

Here you are then - more tales of adventure that millions idly dream of, of a Mediterranean life now nine years long, founded on the rock-bed of love and laughter.

Laugh with me, but know the truth of it.

ROUGH RIDE

To suffering there is a limit; to fearing, none.
Sir Francis Bacon

The Mother's Garden years have galloped away, and with consummate human fallibility I'm forever dropping the reins, wrapping my arms round the neck of life's bronco and hanging on for dear life. But 2003 was the most alarming ride of them all. After we'd whizzed past the post that bitterly cold December I didn't dismount, I slid off.

In the spring Joe Joe's health had stuttered, then nose-dived with chronic pneumonia. We sat in intensive care wondering what on earth we had done bringing our toddler so far from the familiar.

In the bed next to him, beneath the wall of monitors, leads and tubes, was a little girl who had been trying to retrieve a ball from under a car as it was driven away. Her Andalucian father and I couldn't express our feelings to one another (I was trying to get to grips with Catalan at the time, not Spanish), but we sat quietly together for hours, changed into and out of our green masks and gowns together, and sought each other out in the crowded canteen to stir our coffees until they were cold. In the intensive care ward he would lean forward and whisper to his daughter for every second that he was allowed to be with her. The first time I saw her, her face was so badly bruised there was no way of knowing if she could hear him, until she lifted her hand to his face and felt his tears. Joe Joe was nearly three.

We had been in and out of surgeries and hospitals trying to understand how we could fight this. We cut out all dairy produce from his diet and read and listened to all advice going, but every time the weather turned so his chest would begin to rasp, and we would have to race to the doctor. Only that spring the infection seemed so much more virulent, persistent. Seeing him lying there next to that little girl, watching his pained breaths, was the lowest point of this life.

Joe Joe rallied, thank God, Buddha, Allah and all, and a few days before he was finally discharged the little girl from intensive care came trotting through to his hospital bed with a teddy under her arm to say goodbye. Her face still told of the trauma, but she was on her way home first. Her gentle dad hung back at the door to Joe Joe's room, her case at his feet, a smile shining from his tired eyes. I bear-hugged someone whose name I did not know.

So, what of the Spanish health service? Clean, efficient, with the paradox back then of a hospital canteen where, pointlessly, just a handful of passive tables were reserved in a corner for the few of us not trying to smoke ourselves to death. Lighting up in public buildings has since been banned, but this is still a nation with a critical mass of grey-faced puffers. The government is working on it, but you get an idea of how deep-rooted the addiction is when most restaurants and bars, given the option, can't afford to ban smokers.

We have been fortunate to have a 24-hour medical centre three miles from the farm. In an emergency we rarely have to wait to see a doctor or a nurse, and there is also a full-time paediatrician who has become a friend. All fine and dandy, if you can make yourself understood and have enough language to absorb both prognosis and proposed remedy. Through the Joe Joe crisis the only negative, beyond our linguistic fumblings, was how unfriendly the child wards were. The staff were fine, but there was precious little by way of diversions on the walls and windows to counter a young one's fear. Maybe it was because we were in a clinical crisis, but the wards looked like any other, devoid of colour and decoration.

When he was out of intensive care Joe Joe was in a typical two-bed room, where we sat and tried to sleep beside him on a reclining chair. Meals were provided, and one of us was always there to help feed him and to assist with four-hourly horrible medical ordeals of the masks and tubes the little man had to endure. We appreciated that acceptance of us, of the policy of encouraging parental involvement and responsibility, and how they catered for us; but that day they rushed him through to intensive care and blocked our way at the door made us desperate.

Back at the farm, Joe Joe started tearing about with the same volume and velocity, while we looked at our dusty, part-renovated farmhouse and realised it was an impossible place for a child with such a vulnerable chest, unless we made some radical changes. It was a crossroads. We either stayed and fought on, or returned to England. We talked for hours, and decided to stay. Why? It's hard to say now. Maybe it was because we couldn't face retracing our steps, unpicking our lives all over again. Maybe it was because we felt, on balance, the quality of life for our children was still better, and that the climate would help Joe Joe. It was many things. But it was a time of numbness and, for a while, helplessness. We needed our family and friends and it's thanks to them all that we rode out the storm. They were incredible.

We realised that, if nothing else, we desperately needed to do something about the whole first floor where the lime mortar between the old tiles had all but turned to dust, the possible root of illness. We tried to find somewhere else nearby to rent and to bring in reinforcements. But where could we go? Ideally, we wanted to stay in the village one mile away, but it seemed there was nowhere available. The local letting agents had nothing on their books. So I went to the mayor's office. Nuria, the secretary, said she would ask around. We spoke to as many people as we could and within two days we had four offers of accommodation.

Our sleeping quarters for the next six months was to be a little terraced house on, appropriately, St Josep Street, alarmingly close to the church where the bells rattle teeth every quarter of an hour, day and night. We lay there through the long hot summer nights, listening to life on the street, counting the chimes, trying to get some rest as we did battle with the farmhouse. It was, we have deliberated, a very important time for all of us, being in the heart of the village, feeling part of the community, watching the children blend into it all, and we owe a great deal to our friends Remei and Alex for allowing us to rent that little house, and to the accepting and concerned people of St Josep. Like Carlos, for example.

I'd first met Carlos, or Carles in Catalan, about a year before. It had been a bit of a shock for both of us. I was trundling up to the village on some errand or other, the car whining in second gear and my brain in neutral, when I saw Remei's Citroen coming down the hill towards me.

We both slowed so we could pass on the lane. But Remei, beaming her usual broad smile, held up the palm of her hand to stop me.

It was only when the cars were almost level that I noticed there was someone with her. A small bespectacled man being swallowed by the passenger seat was looking up at her open mouthed as she rattled off an explanation of who I was. He seemed to crumple even more as she talked, and I thought I saw water on the fringes of his wide eyes.

"This is Carlos," said Remei after I'd walked over and lent on her door. "He used to live in your house when he was growing up. Others have owned it since, but it was once in his family."

"Hola." I said, offering my hand. Part of my mind was rapidly organising a huge list of questions, while the other half was warning this wasn't a good time. The man was obviously upset.

"His father built the big balsa at the farm. He was there when he was a boy, but Carlos lives near us in the village now."

"Come to the farm whenever you like," I said in Catalan. "Maybe Remei can bring you one day to see it."

Poor Carlos looked bewildered. "Well," Remei interjected, now a little flushed and seeing a need to curtail the conversation. "We must go now."

Shortly afterwards Remei explained what was abundantly clear at the time - that Carlos was very emotional about the farm which he loved so much, and that the surprise meeting with me had been a shock for him. Bumping into the foreigner who had moved into his old home was not something this gentle man was relishing. He was used to a society and a valley that had changed so little over his lifetime and then suddenly there we were, trying for all our worth to blend in but manifestly not of his world or culture. And I was taken aback too, because it was the first time I'd encountered any open unhappiness about our arrival.

When we first moved to Mother's Garden, I'd steeled myself for some open hostility, but there has never really been any. People may have been reserved, or kept their opinions concealed, but as the months passed and we tried to slot into school life and the rhythm, and our contacts broadened, my worries about it eased considerably. So, seeing Carlos and sensing his distress was unnerving. It was arguably worse than if someone had barked at me. Later, Remei said there were bound to be other opportunities to talk once Carlos had got used to the idea, but time slipped by and I didn't see him. It was not until after we'd moved to the village that our paths crossed again.

The front doors of the facing properties in St Josep, like those in all the village streets, are just 12 feet apart, and once the 2003 heat wave had fired up in early June the Latin ritual of sitting outside chatting in the cooler evening air meant it could take quite a while to walk the 20 yards from the car to the front door.

I didn't recognise Carlos at first. The straight-backed, broad smiling man at number five was nothing like the person I remembered from the meeting on the road, and it took several encounters for the penny to drop. Even then I dodged the subject of the farm, fearing he might dissolve again. He always waved a greeting, but there was no attempt at conversation by either of us. Then, one evening in mid June, it was Carlos who broached the subject of *L'Hort de la Mare*, Mother's Garden.

"Wait a moment," he said as I meandered down the road after a particularly difficult day grappling with old wiring. Carlos darted into his door and reappeared a few seconds later holding an old painting of our home.

"This was my bedroom – here – above the kitchen," he said softly, tapping the canvas.

It was fascinating. The house was the same, but the scene in front of it was so changed. Instead of the fat fig trees that now cloud the vistas and cast welcome shade across the red soil there was a clear view of the mountain that watches over the valley. And out from the front wall, beneath the sundial, there stretched a vast pergola that, when covered in vines, would have provided the vital midsummer relief. The timbers looked solid enough to last more than one generation yet we'd found no evidence of them.

Carlos turned the painting over and showed me some writing on the back. "It was painted by the village doctor in 1951. See? Please take the painting for a while. Please."

The painting hung in the broom-cupboard-sized lounge of the rented house for three weeks, and we thought about what life at Mother's Garden must have been like 58 years ago when Franco's Spain was all but closed to the outside world.

As for the repairs, things went from bad to bloody awful.

First we discovered that the beams in the former kitchen-turned-junk-room, supposedly supporting the shower room and main bedroom, complete with its uneven brick floor, were riddled with termites. The wood just crumbled in your hand. How in hell's teeth the whole lot hadn't come crashing down when my visiting 84-year-old father had rolled over in bed we'll never know. So, square one was to go back to bare walls at that end of the house, roof to floor. After that all the rest of the upstairs floor had to be lifted and tons of debris carted out before a new oak floor could be laid. And when we began that battle the ceiling in our farmhouse kitchen, the only room we had completed, caved in.

That might well have finished us, but for our friends.

Mac and Conxita from across the valley, immediately, typically, joined the fray; an English builder friend brought a bricklayer and a carpenter for a week's hard-labour, and – we still thank our lucky stars – that spring we had agreed to let two strangers come and camp on the farm in return for helping out. Their names are Jayne and Adrian, two gentle, special souls who, like Mac and Conxita, were like family to us. They stayed through that year, supporting us both physically and emotionally and we shared a great deal.

Do you believe in fate? There are times when you wonder who is pulling the strings. Adrian and I, it turned out during a tea-break, were born on the same day. I mean the same day, same year, September 21, 1958. How many people do you bump into in this life who came into the world at exactly the same time as you? I rushed into the barn and despite the upheaval managed to lay my hands on my 1958 birthday copy of the long defunct Sunday Graphic. Adrian buried his nose in it for a while then came up for air to share something that nipped in the bud any notion that V-neck soccer shirts and baggy Stanley Matthews shorts were long before our time.

"Listen to this," he offered. "Sports section editorial comment - 'Not far short of £45,000 changed hands three days ago to give Manchester United the right to employ twenty-five-year-old Albert Quixall as a professional footballer. That's inflation. And it means soccer is losing its sense of proportion. Football could spend its money more wisely than paying fancy prices for its players'."

That's how old we are.

Amid the chaos and upheaval there were some vital, enriching moments during that emotional year. Like that one day in baking June when the rigours of the stifling summer heat were forgotten as we stood in a cavern high in the mountains and witnessed Catalan Gerard marry his English bride Rachel.

In the soft light of the last embers of that day, I sat and watched as some of the wedding guests drifted off home beneath a pastel sky alive with blue-black swifts that tore between buildings and soared above the higgledy-piggledy terracotta roof tiles. Villagers stood or sat on the steps of the nearby bar and watched as waiters and kitchen staff bustled about packing away the trappings and leftovers of a veritable feast. The vast swathe of green shading that had been strung above the sports area to spare us the fierce sun was finally being stirred by a gentle, refreshing breeze, and beneath it several guests who still had a little energy left were circled around bridesmaid Ella, who was leading them through some Sevillana dance moves.

Maggie was with her, but there was no way I could join them because Joe Joe, the ring bearer, was asleep in my arms, job done, tummy full. The cocktail of heat, relief, cava and rich cuisine had carried me up some stone steps to a seat with a wonderful view of everything, and into that dream-like space where, despite the blaring music, all seemed so peaceful.

There was also the small matter of my shoes. I'd done my best to look dapper that morning, ironing creases into my fifteen-year-old suit and wiping clean the black shoes Maggie had found for me at a second-hand shop in Barcelona. But barely had the reception got under way (and shortly after Joe Joe had seen fit to save a very gooey sweet for later by sticking it to my trouser leg,) when my left shoe gaped open at the toe in a Burlington Bertie manner. Walking, let alone dancing, was a challenge.

But it didn't matter. The day had been a joy – Rachel and Gerard giving our children starring roles of their own in the ceremony. The cool cavern was the perfect place because, strange as it may sound, it had brought the happy couple together in the first place.

During the last months of the bloody Spanish Civil War in 1938, when Franco's Fascist forces were closing in on the Republican bastion of Barcelona, the cavern had been a hospital staffed in part by British and other foreign doctors and nurses. They were among the thousands of volunteers from abroad who'd felt compelled to join arms in the struggle against Fascism or believed it was morally indefensible not to commit their medical skills to easing the appalling suffering they felt was being ignored by their own governments.

Rachel, from Lancashire but who'd lived in Barcelona for several years and had illustrated a book about the civil war, was an interpreter during a commemoration in the cavern at the end of 2001, when several British veterans of the conflict made a very emotional return. We'd bumped into her in a restaurant that weekend and she'd invited us to attend the commemoration at a village high in the mountains where, it transpired, a local olive oil expert by the name of Gerard was involved in the arrangements. Such is the course of love.

mother's garden

DESPERANTO

The secrecy of my work prevents me from knowing what I am doing.
                    Quarterrmaster Kenwyn Pearson

If I achieve nothing else from this mystery tour, I might just manage to squeeze some choice Catalan expressions into the English language that will freshen it up considerably. They all work very well for me, and are now ingrained vocabulary, although perhaps my usage and definitions might not bare close scrutiny by a Catalan linguist.

Go on, give them a whirl.

OP-PAAAH! (*interjection, noun*); exclamation, with a highly effective stress-defusing quality. Ideal for everything from minor disasters (like dropping an egg) to serious situations (dropping a beehive). Widely used to mask pain and embarrassment, for example when falling out of olive tree. Also a useful noun when all words fail in describing any of the above to a third person.

OYOYOYYYYY! (*interjection*); retort used by third person on hearing about an OP-PAAAH. Important not to get the two confused.

PIM PAM (*adverb*); quick, a short space of time. Non-specific response widely used by builders, car mechanics etc to specific question about how long a task will take.

PIM PAM POM (*adverb*); very quick. Used widely in response to the question "Yes, but when you say pim pam, how quickly do you actually mean?"

SOGRA (*noun*); mother-in-law. Use with care, as not always appreciated by the person it refers to.

FRUIT SECS (*collective noun*); dried fruit: May be offered for dessert. Try not to pull a Benny Hill expression or wet yourself giggling.

Now, hold out the flat of your hand like you are receiving a sixpence from your grandmother, then move it rapidly from left to right a couple of inches both ways. This gesture denotes, it seems, several emotions, chiefly displeasure, and I find it very useful in silently advising my offspring across a crowded room to stop hyperventilating. Some people use it cheekily, but I have stopped trying that after mis-firing with the mayor.

Joe Joe has another Catalan habit that is utterly deflating. If someone, usually me, is saying something stupid he will pucker-up like Marilyn Monroe then tut. Anyone on the receiving end of such a dismissal immediately feels a complete berk.

He also will reinforce angry words by throwing his flat hand passed his ear, a classic Latin gesticulation that's popular, pointlessly, with people on the telephone.

And here's one you best avoid. It is normal to acknowledge someone on the street or in a passing car with a flick of the head, not a nod – jerking the chin upwards like football thug spoiling for a scrap. It is dangerously habit forming.

Forgive me labouring the point, but a foreign language is a formidable challenge for anyone with a middle-aged head already filled with nonsense. The sensations of uselessness, embarrassment and ridicule await you, and if your eyes are not wide enough to see you are about to put your brain and pride on the rack, then think seriously about what you are doing.

By the third year I was feeling pretty confident, but truthfully, was a loose cannon.

"What's needed," I said assuredly in Catalan to Jordi, the carpenter laying the new wooden floor in the farmhouse, as we pondered how to refit the staircase to the mezzanine in Ella's room, "is for us to replace these nails with rabbits."

"Rabbits?" Wild-haired Jordi's eye was twitching.

"Yep, rabbits," I reiterated, revelling in a moment of shoulder-to-shoulder craftsmanship. "Rabbits will make the whole thing a lot more secure."

The word I was looking for, of course, was snails not rabbits (*cargols* not *conills*), my confusion stemming from the fact that screws and snails have the same name, logically if you think about spirals, and that both words began with c. Bloody hell.

The truth is I have not had one formal lesson in Catalan, having learned it on the street, from a friend, from books and on the kitchen table doing homework with the children. Mistake. Sign up for a language course the second you land. I'm pretty solid now and try and learn something new every day, but there has been a litany of embarrassments.

I came home from the school run to find the then mayor Quico (as in Keeko) sitting in our kitchen drinking coffee with Maggie.

He smiled warmly and asked if I'd do something for him. "Anything" has always been my stock reply given that a) it's a golden rule to keep on the right side of a mayor, and b) I like him. Besides, his requests had never amounted to anything too tricky.

"Splendid." Then he rattled off in Catalan - "I've got to go on Radio Catalunya and they want me to bring someone from the village who is from somewhere else, if you know what I mean."

"No I don't."

"They want to feature our village. It's a sort of quiz, an hour-long programme every Saturday morning. The listeners have to guess the village, then they have to guess where you're from. There are clues of course. It's really fun and will be very good for the village. It's a popular show."

"Me?"

"There isn't anyone else."

I glanced at Maggie who whistled silently and looked out of the window. "All in Catalan?"

That was a thought to myself that sort of just came me out, but Quico took it as a dumb inquiry and spoke slowly like I was thick or something.

"Si, of course. It's Radio Cat-al-un-ya."

"A whole hour?"

"With some music. Don't worry, you'll be fine. You only have to talk for a little while. It's mostly about the village."

He looked me straight in the eye in one of those brain freeze moments when nobody can help you. I remember hearing my mouth say "Um, OK". He gave me some paper with information about the show, saying the producer would ring me in the next couple of days to ask me some questions. Then he was gone. I panicked, and immediately made an appointment to see my unofficial language coach Cristina, who runs the local tourist office. She was somewhat tickled by my predicament which didn't help, but we came up with some handy Blairite-type sound-bytes.

The producer duly rang and we discussed suitable clues about Norwich (Maggie and I having agreed earlier that anywhere smaller, like my twin North Norfolk home towns of Holt and Sheringham might be a bit of a tall order for the listeners to identify). Then I was asked to choose a piece of music. "Paperback Writer, the Beatles," I replied as another great anxiety bubbled to the surface. My latest book, a novel, had been on my publisher's desk for an inordinately long time, and I'd been pacing up and down waiting to hear if a year of toil had been for nought.

Anyway, there I was on the train, completely engrossed as Quico and I talked and talked about his life, his village and stuff, and what he thought about the changes in the community. I'd fretted about that long journey, about struggling to find something to say and how the blood was certain to drain from Quico's face with the realisation of just how wobbly my Catalan actually was. But instead the time flew, and he told me so much. Like the story of the windowless building that stands neglected on a terrace between the school and the backs of the terraced houses of the main street. Someone had once commented, ages ago, that it had been the old cinema, but was now used for storage. I would sit looking at it, waiting for the children to spill out of the school gates, wondering what it must be like inside, if it could be put to some use again.

Quico confirmed its past, and said that far from being trashed it still had all its fittings.

"Really?"

"I will find the key and show you inside some time. There is a wonderful wooden floor, and seats and balconies. Things are put away in there now, but if we can ever find the money we would like to try and link it in to the school."

I thought of the engrossing, beautiful Italian film *Cinema Paradiso*, about a little Sicilian boy's wonder at the cinema in his little village and his friendship with the projectionist. I asked Quico if he'd seen it.

"Yes."

"Was it like that? Did the priest vet the films and cut out scenes of people kissing?"

"Sure. And just inside the cinema, in the back row, there were – still are, I'm sure - reserved seats with signs on them for the priest, the mayor and the schoolteacher."

I asked him if he went regularly when he was a boy.

"Of course. I was there a lot. My father was the projectionist."

We also discussed the jaw-dropping plans for twenty-two new homes, what that might mean for a village that has grown so little over the centuries, and how, since we'd come, the number of school children had rocketed from 24 to 50.

As for the radio show, it was a real sweat because they were gabbling so damned fast that I was always several sentences adrift and had to concentrate so hard that just before the end I got cramp in my right leg. It was a dodgy moment because in my pain I leant forward close to the microphone and whispered a naughty Catalan word, one I've unconsciously picked up over the years and which sort of takes the good Lord's name in vain. I'd learned it from countless people who would mutter it in frustration as they'd struggled to get me to understand something.

Thankfully my live broadcast blasphemy was uttered at the same time Quico was in full flight extolling the virtues of the village, and it seemed nobody save the horrified technician beyond the glass picked it up. Just as well. It was the same day that the popular Polish Pope John Paul II was making his way to the pearly gates.

I've done four radio and two TV spots since, probably because of no other reason than comedy, though they have seemed genuinely fascinated as to why anyone should leave a fairly prosperous British existence to sweat on the unforgiving Spanish soil.

It is a tough question with an answer they had difficulty grasping, but I tried to tell them straight. You try getting down to such nitty gritty stuff in a foreign language.

If I'd had the language I would have gone off on one – saying we are losing the plot, telling them to read the likes of Lilias Rider Haggard, who is as relevant today as she was in the 1940s. The daughter of the prolific adventure novelist Sir Henry Rider Haggard (*She, King Solomon's Mines*) was a glorious writer in her own right; mostly natural history, but with sense and the gift to express it.

"We, as well as those we call our enemies, have lost our arcady, and life for many has become mere endurance. We held that it was impossible that men and women should

be content to live from the cradle to the grave amidst the simple things of a natural life. We were bewitched by the gods of speed and luxury..."

That is one of the greatest frustrations all immigrants face – the lack of fluency to express themselves with all the armoury of their mother tongue.

My earlier book about life here has been translated into Catalan, which was worrying for a while given the people named within it, but I got away with nothing more than a mauling by the Barcelona media. My publishers thought it would be a good idea to make a bit of a fuss for the book launch, including a little press gathering and also a radio advert read by me. I wasn't so sure about that.

"It should only take a few minutes," Xavier from the public relations department had said. "Just a short radio advert – fifty two words. You read, and then you go home."

I believed him.

Suddenly I was in a buzzy world of mirrors, striped shirts and braces, packed into a lift with bright young(ish) media things ever conscious of their reflection, and rising past newspaper and magazine offices to radio studios on the fifteenth floor.

My round-shouldered, un-ironed persona made me feel as out of place as Woody Allen at a Beckham Palace drinks party.

The views of the city from the fifteenth floor were impressive, I'll give them that. But everything that ensued wasn't. I was handed the script and ushered into an alarmingly hot and airless recording studio. I did some linguistic limbering up, put on the earphones and then waited for the three people on the other side of the screen to give me a wave.

They were glued to screens, dials and pieces of paper like NASA space scientists. Finally a nod and off I launched. Not bad, I thought. Too long, they thought. Do it faster.

I tried but my pronunciation went, literally, to pot. The head shaking and heated discussions beyond the sound-proof glass told me it wasn't good enough long before one of them switched on the microphone and said it wasn't good enough.

Then they started to rewrite the script. Fourteen attempts later and my piece of paper looked like a sketch of Spaghetti Junction, with whole sentences cut in half and then transposed. My Catalan collapsed, which didn't help the disposition of the woman in charge who kept coaching me with hand signals like a conductor wincing at an unreliable string section. Somehow, in their head-slapping distress, the three forgot I could see them. We may have cut the text from fifty two words to thirty six, but there was no way I was going to repeat anything half as good as the first attempt.

One sticking point was, it turned out, my lousy pronunciation of a single word. The book is entitled *Es Pot Beure Te Amb Porrò?* – a clever play on words but takes a little explaining. Have you ever been hoodwinked during a Costa Brava or Costa Dorada holiday into thinking that it would be fun to try and drink from one of those decanters with a spout, holding it high above your head and trying to pour wine into your mouth? Bang goes your best shirt. Well that is a *porrò*, and the basic book title translation is *Could you drink tea with one of these?*

Fair enough, only if you drop the accent on the last letter of *porrò*, and don't stress it, the word means joint, as in cannabis, something I unwittingly did for several weeks. I wondered why people kept nudging and winking or declaring it was an extremely risqué title.

As for meeting the press, things turned even more nasty. I had to get up at the crack of sparrowfart to catch a train to Barcelona and then make my way to a restaurant where the lure for the journalists was a full English breakfast. Not that I got to sample any. There were ten reporters, five photographers, three publishing minders and a rank of microphones which had me fleeing to the loo. I stumbled through the interviews, choosing my words very carefully, or so I thought, although one reviewer said in her article I was suspiciously cagey. A split second after the grilling was over one of the photographers was leading me out on to the street.

"I've got the perfect idea for the picture," he said and handed me a *porrò* brimming with cold tea. He made me drink from it five times. Five photographers later and I was in serious diarrhoeal crisis. They'd all had the same idea and were equally dedicated to getting several shots in the can; as was I.

mother's garden

THE WILD BOAR REPEATS

You can't depend on your eyes when your imagination is out of focus.

Mark Twain

How do you feel about several hairy brutes with tusks, weighing twice as much as you, setting dogs howling all along the valley by ploughing up your vegetable patch at 2am?

As you lie in the dark being elbowed by your wife the dilemma is – do you let the wild boar demolish the lot, or do you get out there and try and somehow save a few of the lettuces, tomatoes, onions and spinach plants you have slavishly been watering and weeding through the baking summer? I repeat – hairy, tusked, twice as heavy as you, and out there somewhere in the spooky night. Cute and with stripes when small they may be, like the one on this book cover, but I'm not ashamed to say that the stealthy, nocturnal, grunting adults give me the heebeegeebees.

I know, I know – they are intelligent omnivores with fluffy ears, renowned for the loving, protective care of their young, living in large gregarious groups (called sounders, in case you didn't know), shyly and wisely steering clear of the modern day likes of Asterix and Oberlix, several of whom live in our village. Which rather begs the question, how on earth do they manage to survive and multiply despite the popularity of gun and dog hunts from here to Asia? It is a measure of their gumption, and now that they are loose once more in the south of animal-loving Britain one wonders how long it will be before Tyneside allotment holders will be holding emergency meetings.

I'd love to observe them and learn about them, but it's like living with a herd of poltergeists. All you have to go on are fleeting glimpses, hoof marks, ploughed-up land and the hunters' grim trophies and photographs.

After our first year I was beginning to doubt if the boar existed, because I'd seen neither hide nor hair of them at Mother's Garden. This comment tickled Cuban cigar-smoking Enric, a square-shouldered, slightly bow-legged grandfather farmer of scrum-half proportions with forearms like Popeye, whose dining room is dominated by a trophy of one particularly ugly beast. Every time I pass his house I invariably glance up through the open window at the huge hairy head of the wild-eyed creature with its sharp tusks and long snout that seems to reach half way into the room. They can't be that big surely? Oh yes they can...

Come our second, far drier, summer and we (yours truly and the phantom) finally met. Just after midnight I was weaving home along the lane after running our friend Jane back to her house up the valley. Our little Citroen diesel was chugging along when a boar suddenly burst out of the hazels on the left, and stood stock still in front of the car. I slammed on the brakes, shouting in shock, and came to a halt about four feet from it. Judging by its size – about three feet tall and four feet long – and its grubby curled tusks, it had to be a mature male that probably (a quick search on the internet later revealed) tipped the scales around nineteen stone. We looked at each other for several seconds and I got the distinct impression he was chewing over the option of having a tilt at the car. Its head and shoulders were higher than the bonnet and its coarse brown fur reached to the top of its two pointed ears.

He then turned and trotted down the road. I followed, gingerly, still muttering in disbelief and absorbed by the spectacle of something so huge moving so daintily. It stopped again for a second, gave me the beady eye and was gone. What if I'd met one when I used to walk home at 11pm from the village after giving my English class? OYOYOYYYYY!

It was an omen. The next even drier spring and summer we were inundated. There were many village tales of horticultural mayhem as the spirits of the forest came lower and lower looking for sustenance. Most families have strips of land where they tend luscious gardens fed from ancient water courses, and some of the old boys were driven to distraction by almost nightly devastation, all patterned by tell-tale cloven hooves ranging from hefty males to dinky boarlets.

With virtually no rain through that winter, the boar were about from the April, and we first saw wiggly furrows higher on the land where, with unbelievable power, they had carved up the rock-hard earth beneath the almond trees in search of roots, old nuts and worms. Lower and lower they came each night and we had to do something quick to protect the vegetables just outside the back of the house.

We couldn't run to an electric fence, so I had to come up with some other device as ingenious and low cost as my celebrated fig-tickler (a pond net with a bent end tied to a long cane) which enabled us to harvest the ripe fruit before they fell and splattered on the ground.

I set about building a fence of split cane, the principle being that if a boar pushed through it there would be a loud crack which would wake Biba, the only one of our dogs with sufficient gumption to discern real danger from moon shadow. She would then advise the boar of their unpopularity and they would retreat. At very least I would be alerted. This required, of course, bravery on the part of the dog, and obedience on the part of the boar. The flaw in the plan was both parties' intelligence. Biba would bark frantically from a safe distance but was in no way going to grapple with the brutes, and they quickly twigged this. If anyone was going to have to tangle, it would have to be me.

We were not getting much sleep at all, such were the nightly raids and howling going on across the valley, and any fear I had about going out there was rapidly eroded by tiredness and frustration. My first idea was to lean out of the upstairs hall window, shine a torch on to the garden and yell, but although I heard all manner of destruction I never saw a boar. So, one particularly bad night when Biba was beside herself, I got dressed and ventured out.

My first sighting was under the hazel next to the washing line, about 30 feet from the back door. It was a heavily pregnant female with fluffy ears, looking just like a pot-bellied pig in the torchlight, snuffling about completely unbothered by the light or me. I took several steps closer and she vanished into the night.

The second face to face was not so sweet. Biba had stopped barking, and when I got round to the kennel she was sitting like a statue, staring at the vegetable patch. I could see why. The male boar standing beside the melons and dribbling the remains of a tomato plant was seriously ugly.

My forays outside had sapped the torch, but I could still see him – high at the shoulder and maybe almost full grown, with rough fur standing up on his back. Not like the monster that had trotted out in front of my car, but big enough.

Biba slowly turned her head as much to say "Don't look at me, pal".

I moved a little closer until I was next to the dog. Far enough. The beast stood stock still too, waiting. I put the torch under my arm and clapped my hands as hard as I could. The boar went like a bullet, straight through my bamboo and string barrier as if it wasn't there. Later, in daylight, I could see its hoof prints, one and a half inches across and twice the size of any that had been there before.

I wasn't sure what to do. My bamboo alarm system had obviously worked, but it was half wrecked, and if Biba was going to refrain from shouting at anything that big I'd have to think again. So I fired up Nell our old tractor, who lives in a shed ten feet from the tomato plants, turned her round so she faced the garden, left the keys in and rubbed my palms. This was war.

The next night when all hell broke loose I leapt out of bed and in a fizz of adrenalin skipped the lengthy process of getting dressed and just threw on my dressing down and sandals.

I tiptoed round through the moonless blackness to the dogs, felt my way to the tractor and climbed aboard, sitting down very gingerly so the springy seat didn't emit its usual squeak. Away to my right, under the hazels where I'd seen the female, it sounded like there was quite a few of them, but I'd left the torch inside. The plan was to wait until I thought they were in or near the vegetables, then turn on the tractor lights, fire her up and, well, try and scare them. Seriously complicated plan, eh?

I waited and waited. I could hear them, but without moonshine I couldn't be sure where they were. The mosquitoes were homing in on my ankles so I decided to go for it.

BAM. On went the lights. Nothing there. I started the engine, jammed her into gear and roared off, turning the lights onto the hazels. Still nothing. I was committed then and so, naked but for my dressing gown billowing behind, I charged up the land, heading towards the spring where I'd found countless hoof marks on the bank leading to our neighbour's vineyard. I could see the boars' broad backsides disappearing into the night and put my foot down. Unfortunately I was trying to I retie my dressing gown at the same time, took my eye off the land ahead and got a front wheel stuck in a ditch I'd dug while looking for old drainage pipes. Nell stalled. She and I were stuck. I sat there in the dark, took a deep breath and then legged it back to the house like Eric Liddell in Chariots of Fire.

It's just after six in the morning as I write this, and for sure the soft earth soaked by a day of rain will show the comings and goings of boar, large and small. Most of the time the parched ground leaves no clues, but it is April, and a particularly moist one at that. Just in front of the house, thirty metres from our bedroom window, we have created a slipway down to a small terrace that up until now has been crowded with fennel as high as elephant grass. Thank you, said the boar.

Within days the earth slipway was littered with hoof print tell-tales of a busy highway.

The weird thing about wild boar is, like deer and Father Christmas, they are so secretive, and then suddenly you see one in broad daylight in a shopping centre. A boar walked through the middle of the village that dry summer, past the church and down the hill just as you like. And on my way to pick the children up for lunch one day, at 1pm, one waited at the roadside until I had passed, then calmly crossed behind me.

My latest sighting was the best. An enormous brute sped across a higher terrace as I was tethering the ponies out one morning. Then, blow me, if it wasn't the same boar, distinctly chestnut, grown but youthful, that zipped through the almond grove that evening as Joe Joe and I were meandering home after visiting neighbours.

From September through to March occasional gunfire may echo through the mountains on Thursdays and Sundays, as a handful of men from the village try and set their sights on these ghosts of the forest. Dogs and old wisdom are used to track them, without, I'm sure, recourse to the alarming array of boar smells and sounds offered on some American websites. I, for one, won't be spending twelve dollars on a 4oz bottle of boar urine to attract dominant males "which is best used with the Grunt Call", a piece of flexible pipe that emits either sow grunts to attract other hogs, or the distress calls of larger boars in a fight, which must take considerable practice. What I need is a 4oz bottle of hunters' urine and a recording of men in American army fatigues farting, spitting tobacco and shouting "varmints and critters!"

Once every few years our non-spitting local hunters invite the villagers to share in their spoils, and the 10 or so chequered-shirted fathers and grandfathers cook and serve up a stew to more than two hundred people. At the most recent bash we had cheese, chorizo and olives to kick off, then the boar and potato stew with bread, followed by chocolate sweetbread, all washed down with wine or juice; cost, three euros a head.

"How do you do it for the money?" I asked, my friend Albert, one of the 10 or so locals who shoot.

"It is not about money," he replied. "It is about the village. And the village council helps us meet the cost. Want another helping?"

They came round the tables dishing up from huge steaming pots, and only when everyone was doubly fed and watered did Albert and his friends sit down themselves and eat. Rich stuff, boar meat, and it repeated on me all night, but it was worth it.

All you had to do when you were tucking in was to try and keep your eyes off the large screen on the stage where slides of the chase, slaughter and butchering were shown again and again on a continuous loop. I did watch, though, my eyes held by the honesty of it and the acceptance of everyone there that it was quite natural to know exactly where your meal had come from. Country people are so much wiser about food and more open about the truth of meat; in this case wild meat, in a process so far removed from the modern production line slaughterhouse that, understandably, consumers of tidily packaged choice cuts choose to know nothing of. This candour about life and death comes, of course, from closeness to the earth, to nature. It is a solemn thought how the carnivorous world increasingly shields itself from such reality.

Sticking with alarming creatures, I should also tell you we've seen four tigers, two camels and a water buffalo, all just a few miles from the farm.

When the circus cavalcade rolled into a nearby town, and we watched them erect the blue and white big top on waste land by the sports hall, it was like a throwback to Britain of forty or fifty years ago. It was undeniably intriguing, but also had me in a quandary.

Many of Ella's friends were going, and after some soul-searching, I finally gave in under her incessant sleeve pulling. As an animal lover who even feels uneasy about dropping our children's soft toys into the washing machine and then pegging them out on the line by their ears, I was far from comfortable with the prospect of seeing circus creatures, but the anthropologist in me argued I should. For one thing I'd never seen a travelling circus, and it was all part of this cultural curiosity shop of a life we were living. As for Ella, we discussed how it might be upsetting, and she walked into the big top knowing there was something serious to be learned from the experience.

Two hours later, at dusk, we and about four hundred other people re-emerged as a squall lashed the encampment and men were already beginning to pack up the vast tent. Remarkably there was only to be one performance before the troupe moved on to the next town.

We drove home and talked about the wincing spectacle of the woman who made her living by riding a six feet high unicycle on top of a tiny square table while, I should add, using one foot to flick five cups and saucers which she stacked neatly on her head, before finishing with a flourish by adding a spoon and a lump of sugar.

As for the animals, it was very strange indeed. They all appeared at the beginning of the show and, thankfully, it was over in a flash. Their so-called value in the circus seemed to be more spectacle than any form of submission, with the scents and sounds blurring with the magic of the pool of light and the circle of sawdust. But it was very sad all the same, and I couldn't fathom how they reconciled what they obviously saw as an essential element to the circus experience with the huge expense, risk, responsibility and hard labour of carting such a menagerie around the country.

As for the human performers – nine in all, each doing at least two acts – I have to say they beguiled me with their unquestionable pride and commitment as they juggled, swung above our heads and balanced knives on their foreheads.

There was a gloss of professionalism, but it could neither veil the tattiness nor chase out thoughts of how long this extraordinary way of life, this bygone entertainment, has to run. The ringmaster, resplendent in a bright red Les Dawson frilly dress-shirt, oozed charm and genuinely seemed to believe what he was proclaiming, while the young clown acted the fool with aplomb before packing us off home with some out-of-the-blue pathos. He dragged a chair into the middle of the pool of light, pulled a cloth and a polished tin lid from his jacket and then proceeded to take off his face paint, nose and wig. Finally he stood up, shrugged, and proceeded to sing Imagine well enough to have Lennon smiling down on him.

I still can't get my head around the experience. It was appalling, amazing and nonsensical, yet in the most part very clever.

Then, bizarrely, later that week, Joe Joe chose a very cheap DVD at the supermarket when I wasn't looking - it had horses on the cover - and it turned out to be the Circus epic starring John Wayne and Rita Hayworth, set in Barcelona. Freakily, it featured a remarkably similar tiger cage and act and exactly the same clowning routine.

And if that wasn't enough of a coincidence, a visitor listened to the story then revealed that her great uncle had been a high diver, and that she'd grown up in the Fifties in Sussex very close to the winter grounds of a British touring circus. She could remember hearing the animals, and how her playmates were the children of a Hungarian high wire act.

Among the visitors to the farm has been one Oliver, the son of my father's gardener in Norfolk, a young man with an extraordinary knowledge of nature and reptiles in particular. With some twenty or so snakes of his own, Oliver very calmly and impressively taught us a thing or two about some of the creatures with whom we share the land. He didn't bat an eyelid on finding a viper in a ditch at the top of the vineyard, was delighted to glimpse a golden eagle, and was particularly keen to know more when we told him the valley was one of the last outposts of the very rare Mediterranean tortoise.

"But you won't find one of those on the farm," I said. "I don't know anybody except the experts who has seen one in the wild, and anyway, what few there are supposedly stay up there on the little wooded peak above the village."

I think he went looking all the same, but it was a lost cause, of course. The day after he left Maggie rose at dawn to let our oldest hound Charlie out for his constitutional circuit of the vines and came back to bed. A few minutes later we heard him baying dementedly, like he always does when he'd got some poor creature cornered.

I rolled over and Maggie went to investigate. Within seconds she was shouting up to the window. Radar nose Charlie had found a tortoise just fifteen metres from the house, under a hazel.

It was a large female, at least fifty years old, twenty centimetres long and weighing in at nearly two kilos. There was no way we could let her go because the dogs would scent her, so we settled her in a cardboard box and agreed we would call the village council office after breakfast. Leaflets had been printed and handed round just the year before telling of the importance of the local tortoise population and advising people to report any findings. Julià, the handyman and a tortoise expert, came and started swearing in shock. He'd never seen one so old or large. It was as if Oliver had willed it.

This drama was on the cusp of April, the awakening time, with birds returning from Africa, buds bursting, and a curious beep- beep- beep call in the wee small hours. It sounded like a bus or van reversing, for want of a better description, and for a while we didn't know if the source was mechanical, feathered or furry. Many birders know of a similar puzzle that beset a village in Hampshire in 1980. Residents called out the electricity board because they were getting the same weird noises at night from, they thought, some power lines across the village green, but which turned out to be emitting from a rare scops owl in a horse chestnut tree. In our case it was a walnut tree.

April is also the month when the dazzling bee-eaters announce their return, whistling on the wind and painting their jaw-dropping colours in the sky. They are assuredly among the most beautiful birds in the world, but when you've just been gifted three beehives their gold tarnishes. I reckon it took them about a week to catch on, and then they just spent their lazy days, weighing down the branches of the hazels around the hives and decimating my honey factory.

Every time I scared them away they merely regrouped, told some friends to join the party which featured an hilarious and somewhat puce human, and were back on station before I'd gone a hundred yards. The bees, which normally resembled an air traffic controller's worst nightmare in the afternoon heat, were all but gone.

Clever things bees.

Fearing the worst I togged up and went in search of what honey was left, only to find the hives bursting at the seams with bees keeping their heads down.

If you are squeamish about creepy-crawlies, reptiles and scurrying furries don't, repeat, don't buy a property in rural Spain.

As far as possible the buzz word around here is tolerance, We try and keep our farmhouse a squidge-free zone, except where flies and winged biters are concerned. Live and let live if at all possible is the motto. Oh, I have killed. A viper heading for our back door was not spared when perhaps it could have been, and my tally of mosquito hits must be into the millions by now. But I have bent over backwards to somehow escort a variety of rodents, stingers, beetles, centipedes, two snakes and a somewhat distressed bat off the premises. It eats at my conscience if I fail.

It's human nature to resort to annihilation when we fear, loathe, don't understand, can't be bothered or just don't have the time in our hectic schedule to find another way. Don't you think? WHACK. We, the precious creatures, the gods of disinfectant and convenience, have taken swotting to ballistic levels. All the same, when bees start coming out of your bedroom wall, do you a) run out like your hair is alight and return with a toxic death spray, b) calmly ask them to leave, coax them out of the window and then figure out how to get the queen bee out of the chimney, or c) quietly shut the door, keep a safe distance and call a bee expert to take them away?

We went for answer b to begin with, given that I'm an 'L' plate keeper, reasonably calm on the whole and not allergic to stings. But that didn't work. It was not the swarm's fault that I had left an inviting crack in our chimney. And anyway, mass bee murder would be nuts given that civilisation would go hungry if it wasn't for this immeasurably important insect.

That is not a reference to honey consumption.

One mouthful in three of the foods you eat directly or indirectly depends on pollination by bees. Remember that the next time you start flailing around with a rolled up newspaper. Put it another way, if the bees die off then the planet will become a starving, flowerless hell. They are at the heart of life itself. Man and woman have known that since before the Egyptians collected honey 4000 years ago, but today I think we are more ignorant and desensitized to the natural order than ever before.

Worse case scenario? Einstein said that if bees go, then so do we. And for a variety of reasons, some not yet understood, bees are really struggling right now. They need all the help they can get to keep our world together. There's a buzz in the air and honey still for tea, thank goodness. And as if to acknowledge the wonder of it all, the dusky mountain bathes regularly in nectar light. But while Spain may be Europe's largest producer of honey with more than two million hives, the dying is happening here like everywhere. And my heart sinks when I look across at our neighbour's broken land every spring. I close my eyes and send two wishes into the heavens. The first is for the 70,000 bees that live with us. The second is that in all gardens and on all farms devoid of chemical interference there stands a hive (or five); that, for the wellbeing of all things vital, for our survival and for the sheer majesty of nature, bee wisdom is taught in schools.

Most people I have met along my many miles sort of know that bees are important, if not to what vital degree. Mention that you keep bees and they take two steps back.

Hard as one tries to get across how the number of colonies across the globe continues to plummet at an alarming rate, that horrible fact has no sting among the catalogue of seemingly more pressing materialistic catastrophes the populous is told to worry about. Or so I thought.

I mentioned my neighbour. He's a retired pig farmer with about five acres. Across the lane, twenty five feet from the edge of our wild flower meadow, is the entrance to his land where, every Easter, he trundles back and forth on a toxic tractor annihilating everything. Whatever it is he is spraying the concoction wilts all life within 24 hours.

But never for long. By June his fallow land is always green once more - a forest of stout, grim thistles with a couple of outposts of indefatigable poppies - so he sprays yet again, and tries and fails to plough the debris away. Ugly is too weak an adjective.

It's an utter mess. It's as if he is trapped in cyclical Armageddon and doesn't know what else to do. Can you believe people like that? I despair, because every time he leaves his farm, getting out of his car to chain the gateway, he must look out over our tapestry of flowers to the hives on the far terrace, and still he cannot see. Maybe I should give him some honey and offer to cut grass for him. Either that, or widdle in his tractor tank.

Meanwhile, we work towards having four humming hives. Two are active, and beside the barn I have cobbled another out of the serviceable parts of two wrecks. The fourth? With sizable glee we treated ourselves to a brand spanking new one for our seventeenth wedding anniversary on April 23, 2010. Cost? Just €40 including the twelve wax sheets for the frames. This is a basic pine box, I stress, not your cedar wood English craftsmanship, and it won't last many years, but it is so pleasing to the eye all the same, and we know the design works.

We also bought Joe his first veil, smock and gloves, for the little man is calm, fearless, fascinated and eager. Not that I intend to teach him everything. Like how we fix the wax sheets to the wire, for example.

When I first kitted myself out about six years ago I bought a quaint little brass roller that you heat and run along the wires in the frame. The wax melts and bonds with the wire. But it takes an age, which we don't always have. So we use the decidedly dodgy, do-not-attempt-this-at-home Jaume method. Friend Jaume, who has thirty hives, a man of many parts with an easy smile and boundless patience, pieces together a living growing grapes, driving lorries and selling a little honey and vegetables in his wife's village shop. He helped resolve the bees in the bedroom saga, of which I will relay more anon.

All village families here have their agricultural plots with little buildings, and Jaume's is down by the railway line, a minestrone of agricultural machinations that's always brimming with life and piled high with things that will come in handy one day. My kind of guy.

Old chairs circle a grand barbecue that resembles a fireplace in an old manor house. Ten feet away a hive right next to the track dances with life, at the start of a line of acacia saplings that had self sown by the roadside and Jaume had retrieved with the promise of feasts for the bees. Of the many tree blossoms we have at Mother's Garden – almond, apple, pear, quince, persimmon, cherry, medlar, plum - acacia is a glaring omission, an invaluable and gorgeous flowerer, so to the verge I will go this autumn, fork in hand.

We found Jaume's mum in the hot kitchen preparing lunch, with three large noisy buckets around her ankles. Two had chicks in them of varying sizes, the third was home to three goslings. By the time we had returned from a quick tour of the hives all the birds were out basking in the sunshine and Jaume's father and son had rolled up for their meal. We said we should take our leave, but Jaume wanted to show us how he fixed wax into his hive frames. No time I said. He tutted and guided me into his workshop. He laid a wax panel on to the frame and using a car battery charger sparked heat through the wire. It took about twenty seconds.

Back at Mother's Garden we took our first honey of the year, 17lbs, including one delicious frame where the bees had made comb without the need for wax sheet..

Coming back to my first point about bee classes for all ages, I'm serious. I was also going to say that it's a matter of such international importance that the British Government should act, but that's not necessary. There's someone in high office who has long championed the cause, and across the country it seems more and more people are taking up keeping. I spoke to keepers in Yorkshire and East Anglia, and the picture is positive, with a hum of interest following publicity about the diminishing bee population.

What is critical is that funding for research is not severed during the current whirling of the economic axe. Over to you Vince.

The business secretary Vince Cable has long been wise to the issues.

"I asked parliamentary questions about bees, first of Blair, then of Brown and they ridiculed me, saying that I wanted to spend thousands of pounds of research on bees and how this was typical of Lib Dems wanting to spend money on stupid things. I now realise their ridicule was based on incomprehension."

The really good news is that, as well as a phenomenal network of local associations, Britain has a National Bee Unit, believe it or not, and it has been around for more than sixty years working to combat bee pests and diseases, while promoting training for keepers, organising more than seven hundred events for beekeepers each year. Hope, in other words.

Their Beebase website has how-to-get-started guidance for anyone thinking of dipping their finger in the Winnie the Pooh pot. Then, of course, there are the local groups of beekeepers. Finding someone close to home to help you is invaluable. Here things are somewhat less organized, as you might expect, but no less essential. There are estimated to be more than 700,000 million "domestic" bees in Spain, and the earliest known harvesting of honey by man was near Valencia more than 7000 years ago. In every village there are people with the knowledge, passing it on from generation to generation. Like genius Jaume.

Back to the buzz in the bedroom. We rang Jaume because it was going to need his guile and head for heights. While I was pumping smoke into the chimney cavity, he needed to be up a ladder trying to coax the queen and her entourage into a box.

We failed.

Jaume has 30 hives and a lifetime of experience in apiculture, yet the experience left him baffled too. Oh, the bees left the chimney alright, and gathered at the entrance to the box, so we blocked the hole and left them overnight to settle.

At daybreak they lay on the ground, wiped out, it seemed, by the cold. We scratched our heads and mourned. They were a smaller, black honey bee, maybe African. Then the first rays of sunlight reached them and most began to stir and fly away. Extraordinary.

As for the bee-eaters preying on our happy hives, I don't flap anymore. To see them drink from our round reservoir is the rarest of treats as we clock up more and more wildlife memories, some golden, some blood red.

LIFE AND DEATH

*I like to keep a bottle of stimulant handy in
case I see a snake, which I also keep handy.*
W.C. Fields

We have front row seats in the timeless order of life that goes something like this: The eagles eat the snakes, the snakes eat the rats, mice and the geckos, the geckos eat the mosquitoes and the mosquitoes eat us. Oh, and there are some ghastly little biter flies that lurk in the vineyard and around the corral which, if you fail to spot them on their final approach and they manage to lock on, can leave you with a huge, painful swelling – seriously unfunny if it's on your face.

I'm also now, finally, on first name terms with an old friend, namely *Coluber Viridiflavus*, the European whip snake. Those of you who have read *No Going Back – Journey to Mother's Garden* may recall my first meeting with *Coluber*, although at the time I didn't know his name. Shortly after we'd moved in, I came face to face with a large almost entirely black snake that emerged from behind the washing machine in the barn just as I was kneeling beside it and feeling for the flexible, black, waste water pipe.

I now have the measure of him because one hot April day I nearly trod on him while walking the dogs along one of the overgrown terraces, and for the first time in several encounters he kindly stayed still long enough for me to clearly identify him. He is not dangerous (unless you happen to be a rodent, bird or another snake), likes to climb trees and will happily move into empty buildings, which explains our first encounter. After letting me look at him coiled in the long grass for about five minutes he bolted for the dry stone wall, stretching out to just over a metre.

Spain has thirteen snake varieties of which five are venomous. We share the farm with at least five of them (two of which can give you a nasty turn) which explains the almost daily presence in the skies of the snake-eating raptors. Catalonia is said to have the highest rate of snake-bite deaths in Spain, with mountainous, forest areas, er, like our farm, being the worst. But the figures speak for themselves. Deaths from snake bite in the whole of Europe are estimated at about fifty per year with only three to six in Spain, so the odds are more than 13.3 million to one. Death by bee or wasp sting is more likely, although still minutissimal.

There are the numerous water-loving grass snakes (I even saw one slithering hell for leather down the road in the village), the fifty centimetre grey Lataste's vipers that have been known to sun themselves in our flowerbeds, the Montpelliers which can grow to two metres and though less poisonous than the vipers can be inquisitive and aggressive, the whip snakes and the equally large ladder-back snakes. A ladder-back once obligingly climbed an oak sapling and looked me in the eye. My closest shave was with a Montpellier. I was on my knees, head to the ground, reaching into our top balsa to pull the plug and drain it into the larger reservoir. I didn't see the snake curled in the grass next to my head until Tilly our terrier rolled up. In the shock and kerfuffle I half fell in, but wasn't bitten. The golden rule, of course, is don't go gallivanting around the land in sandals the moment the heat rises (but I still do), and don't move rocks with your bare hands. Always look where you are walking, even close to the house. One August, Joe Joe was zipping about in and out of the farmhouse at high velocity, as is his will, when suddenly the normal riot of noise turned into a blood curdling scream.

"Snake!" he bellowed. "It bit me!"

Parental blood drain.

"W-W-What? Where? What did it look like? Which way did it go?"

What we didn't want to hear was that it was small, grey, with a zigzag pattern on its back. As we looked carefully at his foot, Joe Joe pointed under the outdoor sink beside the back door. "It went under there!"

"Was it small or big?"

"Small."

"Green or grey and black."

"Not green."

Panic stations. Then Joe Joe couldn't remember which foot it had bitten. There wasn't a mark on either of them.

"Are you sure it bit you?"

"Well, I saw it."

Crisis over? Er, no.

If the rest of the information from our then four-year-old son was accurate the snake was still against the back wall of the house behind the mop bucket. And just in case Joe Joe had been bitten, I had to know if it was a viper or not, so I picked up a piece of cane and lifted the bucket. If only I'd had the sense to close the back door first.

Quick as a flash the snake shot past me and instead of heading for open ground disappeared into the house.

Maggie, who can deal with rodents and equally ugly insects – and who showed remarkable calm when confronted by a large hissing Montpellier when walking up to our spring - had understandable difficulty coming to terms with the idea that there was a viper loose in our home. I should point out that just inside the back door is our old kitchen, the one room left to decorate that had in the meantime become our store for all our clutter that we didn't want to put in the barn for fear of rats or snakes nesting in it. We were talking large needle and a stack of clobber...

I was fairly certain the snake hadn't turned right into the hall, so set about blocking off the room from the rest of the house. Then, sweltering in wellies, work jeans and leather gloves, I gingerly began emptying the room of boxes of children's paints, chairs, winter shoes, heaters, rugs, you name it. Several juggernaut beetles – huge but harmless creatures that seem hell bent on moving in with us – popped up and made me jump, but there was no sign of the snake. I stopped briefly to cement up a couple of cracks in the wall just in case it had hidden in there, then continued to lug things outside until there was just a wooden chest left. I took a deep breath, lifted one end, the viper shot out, I dropped the chest and, by sheer fluke, managed to trap it. I thought it was a goner, because when I picked it up by the neck it barely moved, hanging like rope, save the flicker of its tongue. I took it down to the lane and was heading for the ditch that leads down to the river when it suddenly raged, curling its body round my wrist. I shook it off, dropped it on to the tarmac, danced backwards and watched it rocket away into the undergrowth.

"I'm not going round that again," I told Maggie. "Let's keep the doors shut all the time."

Ha Ha. That lasted about a week. Well, we live out as much as in, and the day when we unplug our draughty, aged, double-width front door at the end of winter is a time of joy. We might shut the doors and shutters in the fiercest heat to keep the house cool, but for most of the time the sunshine, mountain air and birdsong slip in through the bead curtains and we can look out at dappled walnut and fig shade on the red earth. Beetles, hairy centipedes, sometimes the poisonous ones, slide in to, but save a fat rat there haven't been any other traumas to make us think of shutting out the world.

Rats are seriously smart, a well-known fact. He or she could have been in the house for hours, days even, dodging the dogs in the kitchen and checking out the layout of the whole house from ground to loft, judging by the few calling cards I found after the event. We suddenly realised we had company in the middle of the night when it scampered repeatedly along the void between our floor and the plasterboard I was putting up in the old kitchen below. At the time it sounded like a plague of rats. As is my way, I'd started a job and then left it, so there was just one piece of plasterboard nailed up between the beams with a gap at the end where the wall curved, creating a cosy hideaway.

It's impossible to lie there listening, of course, so up I rose, and found it clinging to some electrical wiring I had pinned to the beams before finishing the plaster work. It obligingly waited there while I galloped out of the front door to the barn to get my catapult. Whack - I was able hit it and triumphantly chased it out of the back door, round the house, in through the open front door and back into the old kitchen again. This prompted a fit of the Basil Fawltys. I spent the next two hours sealing off the old kitchen yet again, muttering incessantly, and gingerly emptying all the clobber. The rat eventually popped out from behind a box of child paints and there was a frantic five minute chase around the room before I lost sight of it again. I sat down exhausted and tried to work out what I would do if I were the rat. My eyes slowly wandered the room, then rested on the coat rack and the shelf above it lined with hats. At exactly that moment the rat popped its head up out of a colourful felt hat my sister had made and stared wide-eyed at me. Game over. I swept all the hats off in the direction of the back door and it was gone into the night. Sorry, Sis, but now you know why your hat lost its charm.

Other interlopers have included a bat which persisted in doing circuits of the living room for an hour despite every encouragement to leave, and a ghastly beastie making loud clawing noises which turned out to be a giant beetle trapped in the children's upturned drum. The only poisonous black widow spider we have seen was in the mountains, but I lost count long ago of the times we have had to deal with aggressive banded centipedes, that can grow to 9cm long and have a nasty sting. It will eat anything smaller than itself, including lizards. Nice. But guess what – we still haven't seen a scorpion on the farm.

Of course, if you opt to buy an ancient dwelling with walls consisting merely of rocks in-filled with mud, you can rest assured you will never be alone. I accept that.

But the mice... The brain-strain regarding them over the past ten years are turning me grey.

I genuinely thought that after plugging a hole that had appeared in an old stone arch in the store room I had finally cracked the problem. They had tried everything – loose floor tiles, chewing through thin plaster, squeezing through tiny seams in the relatively new upstairs wooden floor – and each time I had fixed it.

I was sure I had them beaten. Those I caught were released far far away in the forest.

Then, just before Christmas, Biba the dog started barking in the night and could be found sitting in the middle of the kitchen, hall or office, head swivelling and eyes scanning the floor in a "which way did it go?" fashion. Calling cards began to appear. I naively hoped, as I always do, that it might be an isolated incident, namely a single mouse that had shot in through a door and was living in a corner somewhere. I set a trap and began tracking the blighter. But he or she seemed to be everywhere and nowhere. The evidence was widespread, but there was no sign of the interloper. I rechecked all the former points of entry and they were still air-tight.

Then I got lucky. Early one wintry morning I came in from the ponies and I was warming my bum against the wood stove while waiting for the kettle to boil when Biba started doing her thing in the hall. My eyes scanned the floor then something struck me as odd. The corner of the batik cloth that conceals our head-high electrical switchboard under the stairs was moving, albeit faintly. Then something dangled for a split second – a tail. I ran to it, lifted the cloth and - nothing. I peered all around the plastic casing and there, bottom left, was a little hole. I opened the switchboard and inside was the rubble and dust of expert tunnelling. This was the wall of the once weak plaster before I had sealed it, a former outside wall as thick as from your fingertip to your elbow.

I spent the rest of the morning gingerly spooning rapid cement into the orifices beneath the fuses like a dentist filling cavities. Only if and when we raise enough money to re-render the outside of the house will the shenanigans cease. Maybe.

My assistant throughout the rat pursuit was our old springer Charlie. His reflexes were shot by then – he was twelve and deaf as a post – but he always loved a bit of sport. We miss him so much. He died at fourteen in 2007, which is not a bad innings for the breed in general and for a hound with a habit of pushing his luck. We did the whole nine yards with dog training classes in England, but the moment he could see a horizon he went for it, across busy roads, getting caught in snares, ponds, you name it. During his last years at Mother's Garden, though, we decided to give him his total freedom, and he took it, either doing his chicken aerobics, bouncing up and down outside the hen run, or letting his nose lead him, with his younger sister Megan dutifully at his heel. They worked as a team (one was deaf as a post, the other half blind), roaming the farm, digging holes, eating anything, chasing chickens, and sleeping for a living, and it was normal for them to be absent for an hour or two when the mood or an odour took them. Biba would sometimes join them on the first leg of the journey, to the edge of our land, but for whatever reason (intelligence probably) she never went further. But for her mother and uncle the fact that it might be blisteringly hot was never likely to be a consideration when you are several studs short of a dog collar. Nose first, think later, if at all. Then, one particularly steamy May day, the two of them didn't come back.

Off they bumbled first thing one Tuesday morning, blithely ignoring the fact that, in human terms, they were ninety one and sixty three years old.

I'd had to track them several times before, through the undergrowth or abandoned almond groves and across the vineyards that border our farm, and I once had to fish them out of an abandoned deep-sided farm irrigation pond. That was the closest call, and we thought they'd learned their lesson. But as the hours slipped by and as the sun began to set we feared the worst – another mindless cooling plunge into one of the many unfenced balsas that pattern the valley, some lost to the wilderness and impossible to see. Or, they had mindlessly taken on some wild boar and lost. The chances of finding them were slim. I used the last of the light to revisit the balsa they had fallen into before, then after a night of little sleep rose to a rare but welcome short blast of heavy rainfall at dawn, and trudged off towards a deep almost impenetrable ravine that they liked to explore and which has more than its fair share of abandoned wells and hazards.

All I found were fresh wild boar hoof prints in the mud and spooky shadows in the abundant cane. My wellies filled with water and I staggered home.

Over a fretful coffee I remembered St Christopher, my patron saint of lost things. (I know, I know, you all swear by St Anthony, but Christopher has always done the business for me ever since, many moons ago, I first mistakenly called on him for help and he showed a particular flair at pointing me in the right direction. His namesake tracked down America, after all.)

The sun had burned through by then, and it was nigh on thirty hours since Charlie and Megan had legged it. After reporting them missing to the Guardia Civil police officer who looked glad of something to do, and getting the village announcer to put out a message on the loud-speaker system, I suddenly had this urge to check out the old abandoned paper mill up the valley. It's far beyond their normal range but I felt compelled. It's a vast stone building, long left to rot, with broken roof and deep cellars. If they had ventured in there was no way out unaided.

They weren't there. I couldn't see them among the debris. But instead of turning for home, I shot up the track on the opposite side of the lane, one I'd never explored before. There were some tatty farm buildings. I pulled up at the entrance, and as I got out of the car I heard Megan's bark from deep within a balsa that lay beyond a broken wall and which couldn't be seen from the track. She and Charlie were shivering in the five feet deep weedy water, barely afloat and only alive because they were supported by a few reed stems growing in a corner.

In I went, but not before I'd launched myself off a two metre high terrace on to some stones and scrub. My ankle hurt like hell, but I thought the cracking sound was a twig not my fibula snapping. Charlie's teeth were rattling. He flopped to the ground as weak as a kitten when I found the strength to push him out and, finally, his day and night water ordeal was over. Megan was quicker to recover, but the old boy looked done for. Wrong. He had another full year in him. Within a couple of days he was barking madly at the chickens, and I finally stopped thinking I had a badly sprained ankle and took myself off to the hospital. The tip of my left fibula was several millimetres adrift of the rest of it.

The dogs' rush of blood to the head coincided with one of our own. Have you tried building a horse corral with your leg in plaster? When I went back to hospital six weeks later to have the cast off it looked disgusting, frankly, with the combined aroma of sweaty foot and pony poo, and the nurse hesitated before touching it. Inside were countless grass seeds and three lollipop sticks I'd used to relieve the itching.

Well, it was a pony corral to be more precise, because in 2006 the Mother's Garden menagerie ballooned with the arrival of a two-tone tubby Shetland called La Petita (Little One) and her dinky month-old foal Remoli, which is Catalan for the curl of hair on her forehead. And at the time of writing, four years on, they are still here, doing sterling work, cutting the grass and fertilizing the vines and olives. Remoli is as round as her mum and trying and failing to grow out of the biting stage. Her penchant for kicking and doing manic circuits round the farm show no sign of abating either.

You don't have to have an equestrian hard hat hanging in your hall to know that ponies, just like horses, spell a lot of... er... um... spade work, and it was agreed the children, the owners, would tackle this. That's all that happened. The agreement was oral not written and went in one ear and out the other. We try and enforce it at weekends and during holidays, but Monday to Friday we big people are the prime carers, which I rather like if the truth be known.

The thunder of hooves is part of this life because Ella and Joe Joe demanded it. They have always had the normal (i.e. almost non-stop) sibling disputes, but there was one fact they agreed on for years; an unwavering wish for a horse.

'We want one, PLEEAASSE", they would chorus relentlessly. Not on your nelly, we replied more times than I choose to remember. "We're not going down that bridle path until you show us you are committed. Riding lessons first, plus hard labour in the stables. Then we'll see."

Beautifully handled, we thought, smugly. A week of wheelbarrow work and the whiff of poo and that could well be the end of it.

"Right," they replied, with a touch of that Whitman steel in their eyes and voices.

I'm inclined to be a little dreamy and unfocussed now and again, and there's a smidgen of that in the kids, for which I'm eternally sorry. But they are most certainly, beautifully, half Whitman as well. My in-laws are of tough farming stock - gentle, kind and wise yet unyieldingly determined, i.e. they get the bit between their teeth there's no telling where it will end.

We knew a few friendly people with horses nearby and could have asked them if the kids could help out in return for riding, but that was not an answer. Skilled tuition with due thought to safety was what we needed, hopefully by someone gentle and kind. All in all a very tall order, because we had been far from happy about the way we'd seen some animals being treated or schooled here, and the harsh bits horses can be forced to have in their mouths.

Then (twinkle twinkle lucky stars) someone we'd met several times when she was giving rides on plump ponies during festivals, declared she was opening a riding school the other side of town. Carmé spent much of her late teens and twenties on the European show-jumping circuit, in the manner of the brilliant Michael Whitaker and Harvey Smith, but without the fame or hand signals. She was kind to her horses. She and her Argentinean partner Juan are now among our closest friends.

Ella and Joe Joe immediately enrolled as some of her first pupils.

Joe Joe is never happier than when leading his pals around the farm or building toy horse corrals in his bedroom, whinnying at the top of his voice and galloping his plastic steeds across the wooden floor. His bedroom has a stable door and his idea of a perfect bedtime book is a catalogue of equestrian breeds. He doesn't run, he canters.

When Ella turned ten we signed her up for a week-long course. The idea was that I and Joe Joe would drop her off at 9am every day, and we would hang around to watch and be around the horses for a while, drifting off home when he'd had enough.

But that was never his idea.

He ran into the stable yard, sporting the broadest of grins, asked Carmé what he'd be doing and never looked back. I was immediately invisible. My gentle efforts to explain to him that he was a year below the age limit for the course fell on deaf ears. I pulled a "what do we do now?" face at Carmé. She shooed me away. She knew Joe Joe well enough, for every time we'd been to a fair in town where she could be found giving pony or cart rides in the market place, we'd always lose him in the crowd only to find him with her.

Everything seemed to be going swimmingly until about an hour later when the little man's steely expression of what I thought was concentration suddenly fractured into tears. He looked at me from atop a pony he had led into the corral and said softly, but very firmly, that he wanted to go home.

That's it, I thought; obsession over. The box of toy horses will gather dust.

Then, the next morning, there came again the sound of a plastic stallion charging across the wooden floor and a little boy whinnying.

I peeped over his stable door. "You OK Joe Joe?"

"Yep." Clopperty clip.

"Do, um, you want to go to see the real horses again today?"

"Yep. NEIGH!"

"Is that a yes or a no?"

"Yep."

He was too busy building the High Chapparal to look at me. I couldn't be sure he was listening. "But – have a good think about that, tiger, because, well, yesterday you were upset and wanted to come home. Remember?"

"'Cos the horse stood on my foot," Clipperty clop, clipperty clop. "Alright now."

Ella rode a tall elegant horse called Izara, Carmé's old show-jumper, while Joe Joe whizzed about on two large ponies, Lucky and Heidi, elbows flapping and calling out HUP HUP as he trotted in circles looking like one of the kids out of Norman Thelwell's cartoon annuals. The first time I witnessed my then five-year-old son calling Juan to help him carefully check and clean out the hooves of a horse he could have walked under stopped me in my tracks.

One of the ponies was a placid pregnant dumpling La Petita, and she was Joe Joe's first responsibility when he started at the stables, so when Carmé said the owners wanted to sell her Maggie and I looked at each other just as we did the day we first saw Mother's Garden. Could we? Should we?

Both our children were born in mid June and we somehow convinced ourselves it would be relatively straight forward to clear an overgrown terrace, build a corral and ship in a mare and foal without the children knowing about it. As for factoring in the feed bills and daily labour of looking after two more rather large animals we touched wood and whistled.

We wandered about the land and decided a bramble-clogged, long-lost hazel terrace behind the meadow was the perfect place; so a digger driver from the village kindly reversed into the undergrowth with his tree-eating machine leaving nothing in his wake but organic mulch, while carefully dodging several firs which would provide shade. As peg-leg me was less use than an shot-putter in a 4x4 relay team, we obviously needed help, and got it in the form of two volunteers from north America who somehow found their way, as people do, to our door.

Becky, from Ontario, Canada, and Carmella, from Washington State, USA, who were in Europe working on organic farms, set up home in the caravan under the walnut tree for a few weeks and saw us through, tending the vines and vegetable patch, sharing lovely times around the kitchen table and joining forces with local friends who came to lend a hand in the corral building and birthday fun.

Corrals cannot be hidden from children who are constantly free-ranging, so we span the yarn that we were going to use it as a holding area for Carmé, so she could stop off when she was leading treks through the forest and vineyards.

The children came home on Joe Joe's birthday and La Petita and little Remoli were waiting for them.

Now we are awaiting the arrival of two new foals. Unplanned, I stress. Definitely unplanned. But after sixteen visits by the randy stallion from up the valley it was inevitable.

Under the heading of Small World, here's one of those spooky coincidences. Becky, who grew up on an island on a lake so far north into the wilds of Canada that it can dip to minus fifty in the winter – she should write a book - had a sister who lived in Brandon, Manitoba, known apparently as Canada's Wheat City. Her mum rang Mother's Garden from there a couple of times.

I handed the phone to Becky and drifted down to the holiday cottage to chat to our guests, two families from Oxfordshire. The conversation came round to roots.

"Debs has an interesting past," someone said.

"My parents are English but moved around a bit and I was brought up in Holland," she explained. "The Dutch translation of your book is very good by the way."

"So do you have Dutch nationality?"

"No. I wasn't born there. I was born somewhere nobody in Europe has heard off. A place called Brandon, Manitoba, in Canada."

Some of my strongest anchors of childhood memory were the heartstring garden burials of beloved pets. It is now much the same for our children, only more vivid, I think, such is the honesty of country living. We have lived here long enough to have animal graves all over the place. Charlie and Megan are just outside the kitchen window beneath a flowerbed in a ring of stones, where a seam of soft red rock was the final challenge he presented me. Fig the cat is on the terrace he used to look down on from his favourite tree, and further from the house are the rocks that mark the last resting places of chickens half-eaten by hounds, and the remains of vermin and reptiles that have come to untimely ends. The children have seen some seriously grisly stuff, but life isn't always pretty, and we think that when balanced with the joys of existence such lessons on the circle of life are essential.

If I sound cold-hearted, I'm not. I wept when the dogs died. Charlie had a stroke, recovered enough to zigzag for several weeks to his favourite places close to the house, rallied for a few months and even went walkabout one last time with his sister Megan, before staggering home smelling to high heaven. Just before he left us I watched him make his way to the indent in the soil where he used to lie beside the vineyard. He lifted his nose into the breeze which fanned his great ears into wings, and he closed his fluttering eyes seemingly deep in thought. I see him there still. Megan, bless her, passed away two years later.

Both our families have always had dogs. We can't live without them. After Megan died, it wasn't a case of if, but when we would get company for Biba

June 18, 2009 was Joe Joe's ninth birthday. We looked at each other and decided to go in search of a terrier-type mutt who would run like the wind with him, love him, love us, chase vermin, guard the farm and gift Biba a new lease of life. We have always had medium to large hounds, so small would make a change.

But where to look? A rescue dog perhaps, so we rang our friendly vet and she started to put the word about. We found out there were three large dog pounds within 20 miles of us and made an appointment to visit the nearest.

Have you ever been to a dog pound? Strewth. Your heart falls out of your chest. The racket is unbearable, the desperation even more so.

We'd figured that the younger the dog the more chance for us to train it and for Biba to bond. But these were all full-grown animals, and mostly on the large size. We were about to flee, guiltily, when one of the English workers there told us they had two puppies that might fit the bill.

How the hell do you choose? How can you part them?

Someone had followed the mother after she had scavenged food, and she had taken them to the pups in a mountain cave. Any of you hard hearted enough not to adopt one or both of them after that? We are fudge. They are now ours. And Tilly and Ted are now part of the menagerie, lobbing puppy eccentricities into the Mother's Garden mix. Like tree climbing.

Tilly has taken the canine penchant for pursuit to an altogether new level, her record currently standing at thirty feet up a cypress. I teetered on a top rung of a ladder, coaxed her into my arms, comforted her, settled her on terra firma and watched her tear round to the front of the house and bounce from a chair into a fig tree.

The Catalan pocket rocket, who weighs little more than a breeze, is never more delirious than when taking the game to a cat or rat which has woefully been advised dogs don't climb trees. Teddy Boy, her brother, remains earth-bound due to lightness between the ears and weight around the midriff. His role in life (when not cracking open nuts for his sister) is to sit, chin up until cat or rat decides to make a run for it: Which all makes life a little uncomfortable for Jess, the long-term fat feline rodent deterrent, who suddenly finds himself in the firing line. I'm having a swell time too, being torn by thorns and claws, rescuing an assortment of the valley's traumatised Toms and Tiddles.

Then there was Friday November 28, 2003, an equally rudderless day full of tears and heartache as Fig, our fearless, utterly adorable tabby, left us in the pit of grief. One minute he was his usual smile-provoking ray of sunshine, bounding effortlessly through his daily routine of staying a yard and countless brain cells ahead of the dogs, snoozing in the sun high on the barn wall, patrolling the farm chasing anything that moved and sliding into the house, seeking a lap and a cuddle. The next, he was a shadow, unable to eat and distressingly weak.

Three trips to the vets and finally we knew. He had an incurable virus, either inherited or caught during a night fight with one of the huge ring-tailed wild cats that live along the fringe of the forest. Such things, along with the peril of the ticks that lurk in the grass and carry disease that will shorten life, are things you may not weigh up before taking on this rustic Mediterranean life with all its other promises. Megan had lost the sight in one eye to it and we spent more than seven hundred Euros treating her. Charlie battled the same disease too.

As for Fig, he came to the farm as a lucky one, sole survivor of a litter attacked by dogs on a neighbour's farm. We'd agreed that the only cat we wanted was one that lived and worked in the barn evicting rats and mice, but Fig rapidly charmed his way onto my kitchen chair and into the heart of our lives. We called him Fig because the trees were his favourite perch when he was a kitten. His contentment was so uplifting. He should have had nine lives but died aged one and a half. We held him as the vet put him to sleep, then we brought him home, buried him between the roots of the fig tree beneath the fallen leaves, and then untied the bamboo wind chimes that hang from it. Blow wind, blow.

PICKLES IN ANY LANGUAGE

*I would be the most content if my children
grew up to be the kind of people who think
decorating consists mostly of building
enough bookshelves.*

Anna Quindlen

Joe Joe and Ella can argue colourfully in three tongues, and do most days, as the wafer-thin line between sibling love and loathing has dictated since the first kids on this earth belted one another.

Mud pies (red soil, impossible stains), all going well, until... "No Joe Joe, that's chocolate."

"Soya sauce."

"Estic bé, estàs malament. Chocolate."

"Soya – oomph – sauce!"

"Choc –OW – late!"

"Give me that spoon ARA."

"Nope."

"ARAAAAAA!. Give it HERE."

"Nope. Argh! MUUUUUM!"

"Mama!"

I have a Christmas card on my desk all year round. It was from Joe Joe when he was five, who used his firm grasp of phonetic Catalan (double ll makes a y sound) to write in English "ai laf llu". It's a heartbeat as loud as artist Ella's spell-binding self-portrait in the bathroom, completed when she was about the same age.

Days are like the clouds, don't you think? Sometimes they are dreamlike with a crimson blessing at dawn or dusk, or bleakness itself and nature's face of rage. They pass, never two alike, like relentless time, frequently with unnerving haste. One dew-kissed, breezy, changeable autumn day in 2007 our Ella took the small yet universally immeasurable step in life from primary to senior school.

Our little girl took a deep breath and swam from pond to lake, en route to ocean; we tensed with that parental drying-throat sense of helplessness, knowing words were fading, hoping that what had been given in early years would stand the course of teenage time. For Ella, by then already eye to eye with her mum, the transition from childhood to middle youth was all the greater, for our choices all those years ago have distanced us as far as we were able from a world which sickeningly sees there's a profit to be had in making little people grow up far too quickly. What is it with governments that put such a ridiculously high value on the economy that children's bedrooms need reinforced floors to take the weight of cosmetics and televisions?

But change came, and for the daughter we brought from England aged five and left in a classroom where she could not comprehend a word (how cruel was that?) it meant leaving that little Catalan school she loved.

Every morning since she moved up, we have to get up an hour earlier so she can catch a coach from the village to the institute, as it is known, the secondary school four kilometres away, where young people from across the mountainous Priorat county gather to grapple with the usual subjects, not least a widening society. At the village school there were just fifty-four children and just two in her year. Now she is in a pool of six hundred students, on the school council and swimming hard.

How important was the last summer before her world began to widen. We tried to figure out what we could do to anchor the moment in all our minds, hers, ours and her little brother Joe Joe's too, so much her closest friend up until then, for so long free as birds on the farm. Madly, we opted to head for the French Alps, skirting the Med via the Camargue and Provence, in search of real white horses, Van Gogh, writer Donald Culross Peattie and some indelible adventures; mad in the sense we only had a beaten up fifteen-year-old, 180,000 miles on the clock, clapped-out Citroen ZX for transport, little money and no accommodation booked.

At the end of our last day of that extraordinary holiday Maggie and I left our books beside the wigwams and went in search of our children. For once they were silent. We found them sitting on a rock close to the tree-house we had slept in for two nights, contentedly drawing the vastness of the wolf forest and high peaks that the day before had been the backdrop to adventure. We were nearly two hours up a hairpin lane into the Alps from the Cote d'Azur, north of Monte Carlo, close to the hamlet of Peira Cava, a cluster of houses in the dense firs where, beside an abandoned ski lift, they have taken the new popular challenge of walking wires between trees to an altogether new level.

Before I had gathered what was going on, we were in safety harnesses and hard hats, being briefed by our guide Julien on the finer points of staying alive. Then we were off, Maggie too, high-wiring it, knees wobbling, ashen faced and trying to keep up with our Ella who never looked back. Joe Joe, meanwhile, was taking on what was billed as a junior course yet, we later discovered, involved walking along wires twenty feet off the ground. I couldn't keep up with Ella. I bailed out half way round the second course. She pressed on, finishing with a ludicrously long, month of Sundays zip wire ride across the Alpine valley hundreds of metres above the tree tops. Think of the most extreme zip wire you have ever seen then quadruple the height.

We had found our way there via Vence, a town inland from Nice I'd visited as a child, that was home back in the 1920s to one of our favourite writers, the American naturalist Donald Culross Peattie.

It didn't rain, but that's not to say it wasn't moist. We rode the fairground big wheel at Antibes one evening, and after a disastrous foray on the bumper boats (as opposed to bumper cars), when Joe Joe steered me straight through a waterfall, I walked dripping and bow-legged back to the car park and, very foolishly, opted to drive naked, apart from a sarong, through Nice, aka Inspector Clouseau, where we promptly got lost in traffic.

It was but yesterday, wasn't it, that Ella toddled up to me in our old Norfolk cottage, just before we moved, and earnestly asked if all horses had cowboys? As I watch her now with her Catalan and Spanish friends, switching between languages, I wonder a lot, naturally, how life would be for her if we had stayed in England. There are still rudderless moments when the children openly discuss what it would have been like if we hadn't moved, or declare truthfully how they wish Nana, Grandpa, aunties and uncles and cousins were closer.

For quite a few Sundays through the spring and summer of 2008 I drove Ella all over Catalonia so she could take part in a television documentary about children from around the world who live in this independently-minded north east corner of Spain, bonding and belonging through the common language of Catalan. Of the thirteen children taking part her closest friends were from Senegal and Sweden. The basic idea seemed sound enough, but at first I couldn't see how they were going to string it out to thirteen programmes. Plus I was frustrated by all the travelling and then having to kick my heels for five hours in a place I didn't know, trying my best to find something interesting to do with Joe Joe without spending too much money, while being well aware there were a host of chores and challenges that needed to be faced at home.

We almost pulled out, but the value of it dawned and I organised myself to fill the waiting time more fruitfully, i.e.

planning beach or mountain adventures or, if Joe Joe elected to stay at home, parking some~~one~~ WHERE peaceful and pressing on with this book.

Ella's new friends were also from Asia, the Middle East and all corners of Europe, and it was so striking to see them all playing together in Catalan. The worth of it was that through play she was seeing and thinking about this life, how other people can differ and share so much, and that we all are, fundamentally, the same. The children were interviewed about what they had seen and learned from each other, and the experience and the need to think in this way will, we hope, have value as Ella enters the maelstrom of middle youth.

I am learning too. One weekend we were in a large town close to Barcelona, visiting the home of the Senegalese family. Seven days before I had sat and talked over lunch as best I could with the open-faced, gentle, dignified couple and their eldest daughter from the predominantly Islamic West African country, a once French colony still tied to Gaul by the rally route that ribbons from Paris to the capital Dakar. Their Catalan home, it turned out, was in a maze of tenement blocks in a quarter peopled predominantly by migrant workers from Africa. It was that other world, of no space and the greatest shadows, of washing dangling from fifth floor windows, of challenges that are only on the fringes of affluent Europe's comprehension; the sort of place that makes you fearful, that thumps you with the truth that many people live like this, or far worse. While I have an unerring faith in human nature that has carried me into some pretty grim places all over the world, it is different when it involves one of your children.

When eventually I found Ella and we drove away I asked her what she had discovered. I said, foolishly, that I could not live like that. Yes, she answered, but some people have no choice. The family's flat, though tiny, was, she added, a lovely neat home, bright with colour. Three weeks later it was our turn. We cleaned like crazy, trying to order our home as Ella wanted her friends to see it – tidy and bright with colour.

When I was a little pickle, hi-tech toys amounted to kaleidoscopes and electric train sets. We coveted Crackerjack pencils and Blue Peter badges. It was the Sixties. I had my sisters hand-me-down Rupert Bear annuals (elephant in tweed suit carol singing, that sort of thing), Look and Learn magazine every week, school on Saturday mornings and a strange feeling of not being told everything for my own good. Childhood still ran until the mid teens and prime ministers smoked pipes. I burst into puberty and dabbled in mild baggy-trousered truculence, after flower power and before punk, between the Brylcreem comb-over and vertical gel eras, when the caveman look was in and comb manufacturers went to the wall, when six-inch high 'pimp' platform shoes in assorted colours looked fetching with kipper ties as big as fists and forearms. Happy days.

Ella's fifteen now, as tall as me (5ft 11in). Our home reverberates ceaselessly with Jonas Brothers, Miley Cyrus and Taylor Swift greatest hits, and what's really scary is I didn't need to check the spelling of any of those names. I even find myself unwittingly humming *Burnin' Up* and *Lovebug*. If I leave my computer unattended it is immediately hijacked for the production of YouTube tributes. The laptop is littered with teen idol photographs to the point that the background photograph of pastoral Scotland is pointless and we are going through colour ink cartridges like teabags. So, I fought back.

When I was your age, I said, it was the groovy Seventies. Ha – that was cool.

Yeah?

You bet.

Like what?

Led Zeppelin, Knebworth Open Air Festival 1979, Stairway to Heaven. Try that for starters. (Huh, that will blow her away).

Ella tapped the keys and within a flash there he was, living legend Robert Plant, on YouTube, hands on hips, big hair and spray on jeans, on stage singing the anthem, with a teenage me somewhere among the 200,000 swaying in the night. 200,000!

Ella rolled her eyes.

There were noise complaints from seven miles away, I grinned. No response.

Right. I shoved her out of my office seat and nose-dived into my musical past, which was a bit disturbing for both of us.

Bowie's attire made her laugh. Lou Reed frightened her. She had a point. But *Transformer* is irrefutably the greatest album ever!

Steely Dan, Yes, Dr Feelgood, Pink Floyd, Thin Lizzy, Hawkwind and Barclay James Harvest flashed before our eyes, and I had to admit it was all bound to look ludicrous to anyone who didn't live through it or dress like that. Maggie's offerings of David Cassidy, Genesis and David Essex didn't help one little bit, though I had a good giggle at some of them. At the last second I sensibly pulled out of typing in The Sex Pistols, *God Save The Queen*, but we are going to have to talk about The Clash, The Pistols and the whole Punk revolution when she's older.

Too late, I tried some ageless James Taylor and (Maggie's suggestion) Carole King on her, but Ella had long given up on us. I don't care. Dr Feelgood's *She Does It Right* is rockin' the house as I sit at the computer right now, and I'm seriously thinking of growing my hair. As for the clothes, I'm still wearing them. The YouTube images of my youth may either be in black and white or washed-out colour, but I'm grateful to the Jonas Brothers for taking me back, in a manner of speaking.

As for Joe Joe, he is still our little tiger, of the age when curiosity grows wings, knees bleed and a boy realises he can easily out-sprint his dad.

For his eighth birthday he decided to be Zorro, and we went to the village to pick up Buzz Lightyear, Rambo and Spiderman. Meanwhile Little Red Riding Hood, Woody the Cowboy and two more superheroes were driven to the pool party by their mums.

There were no bookings for the holiday cottage that June week in 2008, so a vague suggestion that we might invite a few pickles round for a swim swiftly ballooned into a riotous splash. I hid from the hosepipe fight and observed the glee. High School Musical belted out of the CD player but could barely be heard above the clamour, and between long contented smiles and sighs I bit on the conundrums of privilege, influences and what makes for the indelible memory of childhood. Within minutes the costumes were gone, trunks were on and the simple element of water play endured as it has and will forever. By the time the farm was bathed in the sweetness of sunset Joe Joe, his pal Juan and I were wandering through the scents of the weeded top vineyard, skimming the grass borders in search of another kingdom.

Our present to our son was an insect net and magnifying glass, while aunts and uncles sent guidebooks and a science box in which to keep his bits, bobs and bugs.

Joe Joe is a jam jar lad, occasionally asking me to punch air holes into lids. For how long, who knows, but his inquiring mind is educating us all. Do you have the faintest idea, for example, how many kinds of beetle there are? Have a stab.

When I was asked, I suggested between four or five hundred, which struck me as a little bold.

Er... no. The answer is about 350,000, or rather that's just the small number humans have identified. The total could be twenty times greater.

"The insect world is another galaxy, an amazing kingdom we barely know." That is how retired scientist Joe Maddox told it in his soft Alabama sing-song voice, eyes bright, a half smile on his lips. He stayed in our cottage for a fortnight with his wife Janice, fellow entomologist and ex-University of Illinois colleague Mike Irwin and his wife Bonnie.

"You know, humans have identified about a million insects, but we figure there are another nine million we haven't."

They came from America to look for birds, to sample Priorat wines and to catch flies. They walked and talked and told us all that had caught their patient eyes. A pine martin visited them in the garden and Joe stood for hours listening to Sardinian and Cetti's warblers in the hope of a glimpse of a hoopoe. And we talked of their eagerness for change in America, of their belief in their senator, one Barack Obama. This drew from us tentative questions of what it must have been like for a liberal young white person growing up on a farm Alabama, the southern state where many of the major events which defined the modern civil rights movement in America took place during the Fifties and Sixties. Joe's experiences and all their opinions were as compelling and uplifting as their contagious enthusiasm for nature.

Mike, who travels the world bug hunting for science, was bubbling with energy, and declared on his arrival that he hoped to sample the finest wines and to catch one specific species of fly. (More mind-boggling info - there are more than one hundred and twenty thousand identified species of the flippin' things). So he set elaborate traps on the farm, along the valley and even among dunes on the coast - great nets that looked like small tents. He smilingly and patiently explained to our Joe Joe what he was doing and why, and lent him his net.

On the last evening, while Mike bagged up for shipping back to the States thousands of flies including, thankfully, at least one of those he was seeking, Joe showed Joe Joe the little x10 magnifying glass that lives in his pocket.

"You've just got to have one of these, because there is something amazing, truly staggering, which is right under our noses and we don't see it."

Before returning to Illinois both couples made priceless gifts of wisdom and wonder to our son and to us all. Now we will see where it will lead; hopefully to a fantastic journey, brief or long who can tell, yet away from the television, the computer and a world that increasingly doesn't notice the detail, and which so neglects imagination.

Ella and Joe Joe; a prize pair of lovable pickles, and I have been in a pickle or two because of them.

It is an undeniable truth, they will tell you, that I rule our house with a rod of celery. I have been known to bellow like a bull with wind, much good that it does. The rest of the time I have the soporific voice of a Radio Three presenter whose been put out to pasture, devoid of zest, and with neither the authority and brain-tongue coordination to proclaim "that was the Dutch Royal Concertgebouw Orchestra performing Mikhail Ippolitov-Ivanov's Caucasian Sketches" without sounding as if I am breaking some teeth in for a friend.

There are some things I draw the line at, not that anyone listens, and it is not every father who can say that because of his children he has had to sleep with half the women in the village.

It's school's out for the sizzling summer from mid June here - a three-month holiday stretching away to mid September - and like families around the globe we're always in a sweat over how the hell we are going to keep the children amused/entertained/challenged week after week after week. You know - kicking off on week one as brightly as Butlins red coats, despite the certainty you aren't going to make it without blowing a few fuses. Have we got the village swimming pool passes? Check. Is Ella going to have summer dance classes? Check. Can we afford to pay for Joe Joe to go to the play mornings run by students throughout July? Check. Have we got the air tickets for the trip to England for my sister's wedding? Check. And, first of all, what about the impending school parents' group weekend away?

I didn't need it, Maggie certainly didn't need it, but the children never had an inkling of doubt. They were going. I capitulated. Maggie could have a quieter time (one of us had to stay back for he animals), and joining in not ducking out was, after all, our mantra.

Here's my log of two days that will live long in the memory.

June 26. 8.45am.

Panic Stations. We'd overslept. Getting the chores done and making the reasonable 10am rendezvous with the other parents and children outside the village museum suddenly looked like a tall order. I morphed into Corporal Jones; with fair reason, because the only arrangement I had grasped was that I was to meet up and tag onto a convoy going goodness knows where. If we missed the museum rendezvous that would be the end of it.

10.02am. I roared up outside the museum to find we were only the second family to show. Like everything here the schedule started to slip before we'd started, so I flopped onto a bench and waved limply as, one by one, fifteen other family groups casually rolled up over the next half an hour. In the end I figured there must have been about 40 of us – 20 adults and roughly the same number of noisy nippers, but like our hens, they never stayed still long enough for me to count them. Then we were away.

11.40am. We pulled into the remote and ghostly-quiet village where we were billeted. It had been a wiggly procession along mountain lanes, punctuated by screeching halts and children throwing up into the ditch.

Phew, it was scorching. Which house exactly, I asked? There, replied Josep, the computer technician and parents' group treasurer, pointing to a tiny house clinging to the end of the street above a precipice. Remei had wandered over to me and her smile fractured as her eyes followed his pointing finger. She gave me an "oooerrr" glance before we pulled our sleeping bags and clobber out of the cars and snaked slowly to the front door. By the time I've wheezed up the first flight of stairs to the alarmingly basic kitchen and dining area there was already nervous laughter from the top floor. I joined the crush to find that our cosy sleeping arrangements amounted to thirty mattresses in a hot loft with four small windows. The kids, naturally, thought it was fandabbydozy and were bouncing wildly on the single, twenty feet long wooded bunk while the flushed grown-ups tried to unpack in a space that would be a squeeze for half our number. While all this was going on, and some mums were rightly fretting that we might all suffocate before we died of embarrassment, I noticed Ferran grappling with a travel cot for his six-month-old daughter. Suddenly the Sixties plastic sofa with cigarette burns that I'd veered round on the first floor looked hugely attractive.

12.40pm. We sat on the terrace of the restaurant overlooking the village pool and sports area where the boys were, somehow, having a kick-about before their afternoon match against the local team. It is circa ninety-five degrees. A cool beer was going down a treat. Quico, the mayor, was telling me we were going to picnic by a river where the children could swim, when it dawned on me I hadn't bumped into him at the lodging. Come to think of it, none of his family was there.

"We're staying here," He grinned, nodding up at the rooms above the restaurant. "Cheers!"

2.30pm. After another nauseating drive along a lane that weaved in and out of the shade of vast bulbous rocks, the children were just about to go swimming in the river beneath a reservoir dam when Angel asked if I'd like to see a cave in a shoulder-squeezing, I'm-not-taking-no-for-an-answer fashion. His fourteen-year-old son Jaume and Josep the treasurer were either side of him nodding, beaming. "Um..." Then Josep's wife Nuria said she would watch over Ella and Joe Joe, and without another word we were off. Angel, beads of sweat on his temples as we yomped through trees and heat, turned and asked me if I happened to have a sweater and a torch with me. Er, no. We wheezed up to a little bridge where youngsters were plummeting from a perilous height into the water, then he led us along a diminishing path that all but disappeared into the dense undergrowth. It was so hot I could barely breathe when, suddenly, there we were, in a vast cavern standing by the edge of a pool of icy water.

Lovely. Very cold, but tolerable. Very nice, I thought. Worth the struggle. Then I noticed the others were putting on sweatshirts, taking off their socks and trainers, switching on torches and wading into the water towards an alarming small hole at the far end. Josep handed me a little head torch which pegged out the moment I saw his bum disappearing through the gap.

3.15pm. After crawling blindly through water, my head scraping against rock, I emerged damp and cold into a cave with wet slippery walls that look like wax, filled with the echo of running water. It was pitch black but for the narrow, soft beams of light from the other three torches, and the air was bitter. My mind was a muddle of wonder, worry about a freak sudden downpour, and serious doubt that we are not exactly equipped for such a jaunt. But before I could muster any appropriate vocabulary Angel somehow scaled up a slimy slope and disappeared through another tight squeeze.

3.40pm. Three caves and several bruises later there was daylight ahead and we emerged among some dense undergrowth higher up the valley. Angel relayed proudly in machine-gun Catalan as we pulled ourselves up a convenient rope how he'd discovered the caves when he was a lad. A pot-holer, complete with helmet, miner's head torch and all the kit, who was just getting his ropes and act together before descending into the blackness, stood stunned as we emerged looking like we've just wandered off a beach.

5.00pm. Back at the village. The temperature dipped to ninety as our tiny soccer stars were bludgeoned 8-2 by the far larger local youths.

9.00pm. After an undignified quick change in the loft we agreed it would be pleasant to take a walk around the alleys before meandering to the sports area beside the restaurant, where we were invited to join villagers for an outdoor meal.

9.30pm. The huge fish, ham and salad first course was served just as the sunlight faded and Ferran predicted it will only be a matter of minutes before the mosquitoes descended.

9.35pm. I started slapping my ankles and fishing mosquitoes out of my wine.

1.00am. I was replete, mildly squiffy and suitably anaesthetised to both the insect bites and teeth rattling firecrackers being lobbed close to our feet by some of the older children. Joe Joe was spark out on three wooden chairs alongside his schoolmates Ignasi and Bernat, oblivious to the racket and the fact that the cabaret – two crooners on a fairly elaborate stage – had just kicked off.

4.00am. All the children were put to bed. The crooners could still be heard belting out Julio Iglesias's greatest hits. Someone had bagged the plastic sofa. Three of the mums and one dad discussed the merits of sleeping on the street but I tried to settle on the floor at the end of the mattress in the crowded loft, where Joe Joe smiled in his slumber. The steamy room rumbled with snores and the occasional trumpet blast.

6.30am. Joe Joe woke, so we went down to the street, where the four adults were lying in their sleeping bags on the concrete. Joe Joe and I cuddled up on a couple of pillows against the wall, and the next two hours were a sleepless blur, until Ella and the others staggered out of the front door and we joined the painfully slow bathroom queue.

10.30am. Following a creosote coffee and lump of cake at the restaurant we slipped away from the group and I drove the children to a mountain-top hermitage. Joe Joe slept through it, but Ella and I stood and stared at distant Aragon.

2.00pm. Back to the restaurant for an almost silent lunch, when there were sudden howls of laughter. Little Ignasi had fallen asleep with his head in his plate of pasta.

Put a price on that lot. Our bill for the three of us came to just €48 (then the equivalent of about £30) inclusive of two slap-up meals with drinks thrown in, fireworks and concert, a cave excursion and the shock of seeing a significant number of the village mums in their underwear and nighties. As for them seeing my floral boxer shorts and makeshift jimjams (Indian cotton trousers), let's just say everyone was in general agreement. It was an unrepeatable 36 hours.

I am forever trying to pin things - a sight, a sound, a second - on the notice-board of memory. Some fall off, but my habit of scribbling helps things to stay, awhile at least. One morning I rose early, before the sun, with money (or rather the lack of it) on my mind, and made a cup of tea. Our oak kitchen table was once at the heart of my grandparents' home, and I remember sitting at it circa seven years of age, lining up toys and tucking into grandma's yummy-runny mince with peas and mash, always at the end with the candle burn. And there, that early morning, as if it was always meant to be but was just a moment, lays the wooden sword I made for my children. I sat looking at it for ages, sipping Earl Grey.

Being around Joe Joe is something I have known all his life, but not Ella. I missed much of her first five years, and despite the time we have had since coming here, I still feel anxious that I may miss something, and very emotional when I don't.

When she was ten she and I took a stroll. It was in the foothills falling down to the sea, through the manicured, oasis world of retirement Spain. We were drifting through the middle of one of those jaw-dropping, luxury developments springing up out of the arid landscape, just inland from the densely crowded cusp of the Mediterranean where, in the blink of an eye, the stony, parched earth turns into the lush carpet grass of a golf course, neatly adorned with transplanted palm trees, all circling a fake village of plush apartments and houses set along avenues of Cyprus trees.

I was very uneasy. The issues of water shortages and golf course sprinklers gets my gander up so easily. But it was easy to sense the seduction that sucks north Europeans south to such privilege. As it happened, I spoke at the English National Parks Convention two months later on the thorny issue of unsustainable tourism, when I tried to tell Spain's other story, of the little-touched hinterland that few people see.

As Ella and I joined the flow of people, golf buggies ferried the less mobile (or better healed) up the long incline towards the bright lights of our destination, a great stage erected facing the clubhouse. We were there, along with a thousand others, to see Paco de Lucia, Spain's fabuloso flamenco guitarist. The tickets were for Maggie and me, a treat, a respite, but she had had to return to England for a few days, and immediately suggested Ella should go.

Ella looked about her, drinking it all in. The energy of flamenco flowed into the night, moths circled crazily in the beams of light above Paco as, halfway through the concert, Ella lay her head on my lap, smiled up at me and went to sleep.

That same summer a family friend, schoolteacher and artist Eva who had moved to Valencia, whisked her away for four days. Her daughter Julia and Ella had hit it off ,and before we knew it our girl was 250 kilometres away. The idea was that Ella would get the chance to switch her brain from Catalan to Spanish while having some fun. Even so, we half expected the phone to ring and for me to go and fetch her after one or two nights. But she didn't phone.

There was another surprise too. When we asked Eva if Ella was making progress with her Spanish. She looked at us incredulously.

"But she's perfect."

"Perfect?"

"Absolutely. Her Spanish is fantastic."

She is now taller than Maggie, at the end of her third year at the local high school, sailing through most subjects save maths, which is hereditary, is a committed and accomplished artist and a member of one of Catalonia's finest amateur dance companies. Joe Joe, on the other hand, confounds us all with his aptitude and enjoyment of arithmetic, while free-ranging the land, belting his drum kit, re-arranging his countless Playmobil horses and singing Justin Bieber hits.

mother's garden

WATER TORTURE

A fool and water will go the way they are
diverted.

African proverb

We all have our distinct reasons and dreams for doing (or dabbling with the idea of doing) this go-wild-somewhere-warm thing. So those fresh adventurers who beat a path to our door, fizzing with glee and eyebrows-in-orbit adrenalin, looking for answers and pointers, usually find I dry up pretty quick, like most things here. Practical advice I will give, but the whys and wherefores of such fundamental decisions are part of the delicate embroideries of individual lives and impossible for someone else to unpick.

And if they say they have already bought somewhere, in the middle of nowhere "with stunning views and great potential", but with neither a well nor a spring, I glaze.

Property and planning law in Spain can be like reaching on tiptoe for a forbidden blackberry in a bramble tangle. You need a good lawyer with a machete to clear a path or to drag you out by the ankles and tend your wounds, because without expert help you run the risk of being irrecoverably scarred financially and emotionally. The nod and a wink, do-as-you-please days are long gone, yet people still think they can buy a farm building and turn it into a home. You hear of calculated risks, foreigners believing they might just get a fine, or that if the work goes undetected for a couple of years they will be in the clear. Nope.

Then there is the sunset trap. I can name numerous little farms near us with delicious little stone casitas, all legally sound, where it is so warming to imagine oneself living a far simpler life, getting back to the earth, tending the garden, eating well and frugally on home-grown vegetables, having a brood of hens, a faithful dog, and priceless peace: Sitting under a pergola dripping with grapes and quaffing a glass of fine wine as the embers of the day fire the sierra skyline. Absolutely. Pour me a glass.

But... how far from a tarmac road, or civilization for that matter, is it exactly? And no electricity you say. Oh, solar panels and batteries are the plan. Right. Any inkling of the cost? And what about water? No problem, that's good... because there's a huge tank under the house to catch rainwater from the roof? Oh.

Iron grey clouds cast their shadows once in a while, and there may be distant rumbles and great gusts of air to turn over chairs, but that doesn't mean it is going to rain. It's a common tease.

Through all of 2007 and early in 2008, I made detailed studies of those rarities called clouds, and lost count of the times I misread them.

"It's bloody well got to rain today, it has to. Look at how grey and heavy they are. COME ON! Over here you bastards. RAIN, DAMN YOOOOOU!," I raged, like Basil.

And by the clouds would whizz without depositing a drop, until Barcelona was down to its last cupful and our previously constant spring had diminished to a pathetic dribble. The level in the well was falling too, and trees and plants were dying. But we were very lucky. For months those relying on roof rainwater were making almost daily trips down their bumpy tracks to the village spring with enough containers to wash themselves, their clothes, their dishes, to clean their teeth and flush their toilets. Carrying every drop you need makes you realise the weight of life. The old country proverb, that you don't miss water until your well dries up, should be plastered on billboards around a have-it-all world that needs to start giving a damn about water before wars are waged over it.

Our spring flows from somewhere or other, which I figured was all I would ever need to know, though I've often wondered who through the centuries has drunk from it and lived beside it. Our valley, we are told, was popular with the wealthy and powerful in Roman times, and we keep finding tantalizing proof – a coin, some perfect little squares of worn stone.

But after a few years the spring diminished gradually from a steady stream to a dribble. We had been told that it had always poured at a reliable pace into our smaller, fifty cubic metre, reservoir (the size and height of your average double garage) filling it in three days, but this had turned into something that would take a month. Lack of rain was bound to be a factor, but that couldn't be the only reason.

Someone, goodness knows when, had laid a clay pipe and built three inspection holes, rising up the land towards the source. I could figure that much out at least. If you pulled aside the slab of stone covering the top hole, dangled your head and torso in almost to the point of toppling, allowed your eyes time to adjust and dropped a stone you could just make out a a plop and a shimmer. There was, for sure, a pool of indeterminate depth at the base of an alarmingly narrow shaft carved out of the rock.

So, I figured the old clay pipe must be blocked and that there was nothing else for it but to clamber in and out of the other far shallower inspection holes to try to rod through. The problem was, I soon found out, that tree roots had wiggled their way through cracks in the clay pipe and had almost plugged it. For days I pushed and pulled, flushing out tadpoles, frogs and snakes and dragging free root balls as long and flush as a fox's tail. But it wasn't a long-term solution. If I left it at that then the problem would obviously reoccur pretty quickly, so I decided to try and force a fifty millimetre rubber pipe little by little along the fifty metres. For what seemed like an eternity, but was only three painful days, all that was coming out the other end was the echo of me effing and blinding.

Once that was finally done, I expected things might flow, but it wasn't any different, or possibly worse. There was nothing else for it - I'd have to descend into the spring shaft and check it out. Hopefully, it was just a simple case of the pipe I had rammed through not being below the water level.

I should say now that two things that give me the heebie-jeebies are tight squeezes and watching people do perilous things for the hell of it. You won't find me leaning over a bridge cheering a bungee jumper, pawing over pot-holing maps of The Dales, or biding my time beside the open door of a sky-diving Cessna. But this I couldn't dodge. All the sweat and cursing and seventy five Euros-worth of piping would have been for nothing if the end of the pipe was bending upwards for some reason.

First, I lit a piece of paper and dropped it in to check there was oxygen at the bottom. It glowed until it hit the water. Fair enough. My ladder was just about long enough, but it left precious little room for me to squeeze down. I had the option of the old foot holes in the rock, but I couldn't see more than the first two and I would have had to feel my way down. I didn't warm to that idea. With barely enough room to breathe or bend my knees, down the ladder I scraped, twisting any which way I could to keep a foothold.

I splashed into the icy pool at the bottom and my old wellies leaked like sieves. It was the first water contact they had had in more than a year. I found that the gloomy shaft had widened a little and I was able to squat. After my eyes adjusted, I figured out as quickly as possible what I needed to do. Get a torch. What a wassock. I squinted and could just make out that water was still draining into the old clay pipe.

This needed plugging with rags, and it also dawned that I needed a saw or knife to trim the new pipe which extended about a foot into the shaft and defiantly refused to stop bending upwards like a banana.

The face of my friend and neighbour Mac was framed in the square of light above me. He volunteered to go and fetch tools from the barn, but there was no way I was going to stay down there on my tod, so told him I was coming up. Then I looked over my shoulder. A four feet high, almost shoulder-wide hand-chiselled, seriously scary tunnel disappeared into the hillside. Who created it? How? When? And what was at the other end? Was this the water source?

For the next descent, this time with visiting school pal Mike on watching brief, I was armed with saw, torch, rags and camera, and half way down got jammed. It was one of those moments when mirth masks fear in the hope of staving off blind panic. I eventually managed to reverse out, left the tools with Mike to lower down, took a few deep breaths, had a few stern words with myself regarding mind over matter and promised the coward within that just one more descent would do it. I took my boots and wet socks off which strangely helped; something to do with toes, I think, and feeling my way. With the clay pipe plugged, and new pipe sawn to a suitable length, I resisted the urge to bolt, turned and shone the torch along the tunnel. It ran for about fifteen feet, then widened considerably which, when I was half-way along it, I realised had been created by some sort of cave-in. I came out of there with a following wind, and can tell you with all certainty, I won't be doing that ever, no never, again.

What was needed after that heart-racer was some gentle open-air activity – like joining hundreds of other people the next morning to plant trees on the site of the old village tip; happy chat, crisp air and views of distant Aragon.

Nowt dangerous about planting saplings, eh?

"Oh no..." My blood ran cold as a crowd of people suddenly bunched tightly together and up went a human castle.

119

Catalans are famous for this. It could be described as a crazy sport but in truth is much more significant. It is an age-old part of their indefatigable sense of identity. The practice is believed to have originated from human towers which were built by dance groups at the end of 17th and 18th century. Teams train to compete at festivals, where precise techniques are used to build the highest and most complex castle, a feat of not only muscle power, but also teamwork, concentration and discipline.

The motto of the *castellers* is strength, balance, courage and reason. All fine and dandy, but it doesn't mean I have to watch.

The *castell* that rose from the tree-planting crowd was a relatively modest one, but with a small child in a crash helmet on the top nonetheless. They can be anything up to eight levels high with four people on each storey. Some teams have even managed ten storeys, forming a structure reaching up to a height of thirteen metres, but I'm told a *castell* is only considered a success when it has been successfully assembled and dismantled without collapsing. And, yes, they do collapse, although not that Sunday.

It was at that same spot that I had stood a year before and wondered numbly if our world was about to go up in smoke.

The mayor and about fifty villagers were with me, all statues with arms folded, throats dry, shocked into silence as the flaring flames spread, and more and more yellow helicopters with great sacks of water tethered beneath them raced back and forth into the towering grey pall.

Our (and all of Spain's) greatest dread through the tinder dry summers, or any prolonged drought for that matter, is the seemingly unstoppable monster of fire, for every year great swathes of this country are scarred by it, sometimes through nature's hand, sometimes through human stupidity or madness. You will have heard the stories and most likely seen the agonies and destruction on the television.

The rocky wall of the valley that faces Mother's Garden, part stone part trees, is a towering wonder day and night through all seasons, the lip of the great wilderness, a constant reminder of just how close to the edge we are.

Well, at the end of August 2006 it finally happened.

I was bumbling about the house when I suddenly had the sickening sense that something was wrong. It was about noon, and the familiar sound of a helicopter rose in the valley but didn't pass. It drew me out of the house and then I knew. Smoke was billowing into the sky from somewhere in the forest just beyond the ridge, barely a mile from the farm.

Water carrying helicopters are the first line of defence across these inaccessible lands, and we watched, scared to near sickness, as the single machine raced back and forth, sucking up water from a nearby reservoir and dropping it on to the spreading inferno. Every time I turn from the top road into our forested valley it always flashes through my mind how at risk it is, how much it would be changed if disaster struck. The bitter possibility is not if, but when. The only drop of comfort is we are not so distant from the fire-fighters' airport base.

Finally other helicopters came, then flying boats that can carry far more water. We rounded up the children, telling them not to leave the house, and explaining as calmly as possible that they shouldn't be frightened if a helicopter came to draw water from our reservoir or the swimming pool. Then we stood and stared and willed it to end, praying that the frequent afternoon wind didn't rise to fan the flames.

There wasn't, I kept telling myself, any direct threat to the farm. Was there? We have always worked to keep clear land around us, a firewall of sorts to protect dwelling, barn and holiday cottage. But there is no telling with fire, no knowing what might happen. I paced. We would need to have an evacuation plan.

I wondered what could have started it. The boar hunters had been out the day before and the guns had echoed. Smoking is still a great Spanish sickness. It had rained hard just two days before, so maybe one of them thought there was no risk. How foolish.

The pilot of the first helicopter was taking huge risks, dropping out of sight among the trees to draw water from some unseen source, then skimming the tops to get back to the blaze as fast as he or she could until help arrived, both in the air and on the ground. Slowly, after two terrible hours had passed, the smoke began to lessen. They had defeated it. In total only about ten acres were turned to ash. How grateful we were; how fortunate that there was no strong wind to enrage the fire.

Not so long before, about twenty kilometres to the north, a man burning olive prunings without a permit started a blaze that incinerated more than 1000 acres of forest and olive groves. Two years later, about forty miles south of us, five firemen died as gusts fanned a vast inferno that could not be contained for nearly a week.

And the cause of our fire alarm? I owe the hunters an apology.

I said that a few days before the blaze over the ridge it had rained buckets for an hour or so as a summer storm swept in from the west. When they forecast rain here we first saw it as a release from the fear of fire. Rattling thunder is a small price to pay for the refreshing air, the scent of wetness and the replenishing of the well and spring. But there lies the danger. A fireman told us a lightning strike had been to blame.

How can that be, I asked. Yes, there had been a flash and instant clatter right overhead, that made us wonder if the house had been struck. But that was at the beginning of the storm and there had been a prolonged downpour which surely would have put paid to any fire. Not so simple. The lightning strikes a tree, burning it to the core. The rains come, but the core continues to smoulder. The sun and warmth return, the tree falls and suddenly all hell breaks loose.

Pilots patrol almost daily through the dry times of the year, i.e. all the time, and every farm reservoir is visited and mapped by the local firemen. People are posted on peaks to watch when all is tinder dry, while forest rangers enforce the blanket ban on farm fires. You can only burn cuttings in winter without a permit, and all bonfires in spring and autumn require written permissions and must be damped down by noon at the very latest. Everything, it seems, that can be done is done to prepare for the worst, but if fire combines with ceaseless gale it is a battle that cannot be won until nature decides.

I told how we took a break late one summer and headed with the children for the south of France in search of adventure, being unclear on where we would stay or if our clapped out Citroen ZX would make it. Close to Marseilles we witnessed a vast blaze, fanned by the mistral, claiming swathes of coastal woodland and countless homes. From Barcelona to Nice the landscape bore intermittent scars of black ash.

On our return home, as we crossed the pass into our valley, the scent of disaster was still in the air. Hundreds of acres of forest had burned for days on the ridge where fifty wind turbines are lined, just six kilometres from Mother's Garden. It had raged with such intensity and surged so close to a village – barely fifty metres - that families abandoned their homes. Friends caught up in it joined the desperate struggle on the ground to cut breaks in the undergrowth in an attempt to halt it. Barny and Paul had to fell a giant fir close to their house. Paul and another friend, Leo, faces and clothes blackened like sweeps, were on their knees with exhaustion. And, once more, the winds subsided enough, just in time.

Here in the mountains, far from kind sea breezes, the midsummer still air can reach forty degrees Centigrade, one hundred and four degrees Fahrenheit. At times like these you don't care how many frogs, fish and snakes are in the balsa. You learn to rise and work from dawn, to keep the shutters tight all day and try to sleep through the worst of it; or go in search of other places to cool off, like the public open air swimming pools than can be found in every village. Once, to the shock of the families swimming or lying in the shade of the mulberry trees, a thirsty eagle landed among them, walked to the water to drink before flapping heavily into the cloudless blue.

We know some special places, pools and cascades high in the mountains, but even these can run dry. We went one day when summer was sinking her teeth in deeper than was bearable, and the rocks where the water would flow were hot enough to cook on. It was scary. But instead of turning for home we decided to gasp and climb into the high forest in search of the spring source. Then there it was, a diminished but still joyful outpouring of life, of fast flowing, bright, cool water, twisting and turning as it hugged the side of the gorge in an ancient man-made stone gully. We decided to follow it higher, the children splashing along it until I fished out a snake.

The ravine, locked in by pines and evergreen oaks, had dark, brooding caverns in the red rock, and there were the remains of stone terraces and the dry crumbling stumps of ancient olive trees. With its water source and the narrow strip of fertile, flat ground we talked of how this could have been settled as far back as the Stone Age, like the caves close to our farm. For these lands have the handprints of all ages of mankind.

Pere, the blacksmith who made the railings for our holiday house and who gave me a stone axe head he'd found near Mother's Garden, is the village archaeologist, and his home creaks with the weight of history stored in it. I imagined him pottering about up there near that flow, looking for clues, as on we wandered upwards hoping to find the source of the water, until the channel, in an effort to shake us off and to keep its secret, turned across a little aqueduct and vanished into the shadows of the trees.

We turned for home, and back at the farm I could smell burning coming from the barn.

The rags we had used to apply a linseed and white spirit mix to the terracotta floor tiles in the holiday house, and which I'd foolishly left on a shelf next to paints and other chemicals, were smouldering and giving off a pungent grey smoke. I just managed to get them and the half empty, half melted linseed oil containers outside when they burst into flames.

The economic drought, of course, has significantly sapped Spain's tourism like it has all aspects of this young economy. Nevertheless, there continue to be millions of sun and sangria seekers from northern Europe every year, each using up to two hundred litres of water a day. These sorts of numbers are unmanageable for me, but what I find especially confounding at particularly arid times are the sprinklers on the thousands of golf greens that decorate the parched coastline like lily-pads.

Water, or rather the lack of it, is Spain's greatest challenge, for it irrigates the nation's prime source of income.

Soon after we arrived, we marched through the streets of Barcelona and Tortosa against a government scheme to pump water from the nearby river Ebre to tourist centres and cities in the south. The plan was axed when the socialists ousted the conservatives in the after-shock and recriminations over the Madrid train bombings, but until that point the pipeline plan seemed unstoppable. Feelings ran as deep as the great river that runs right across the north of Spain from Cantabria and the Picos de Europa mountains. Among the 500,000 taking part in the Barcelona protest were some who were naked, painted blue like the water that sustains the valley's orange groves and the Ebre Delta, one of Europe's greatest wildlife wetlands.

I didn't appreciate, though, how high the Spanish water shortage of 2007 was on world news agendas until the BBC rang me up for a chat. I don't think I said what they expected, but it was good to have my tuppence worth.

Yes it's tough, I answered. Some areas of Spain are in a fix and people know they must be extra wise. Some trees and other crops will fail and die, some swimming pools will not be filled, but life will go on. Spain is a land of extremes, and these people are accustomed to its character. The challenge is educating the young and those millions seeking respite from damp northern countries where leaving a tap or shower running unnecessarily does not prick the conscience.

A visitor, Manuel, told me of his youth in Andalucía, how they would fill the bath when the taps would work, to save enough to last them through the long hours when the pipes were dry. I humbly suggested to the BBC presenter there was a need for perspective. I'd just heard the latest World Service radio report on Niger where people were dying of thirst. Am I wrong to feel that there is an alarming resistance to news from the third world because it is irrelevant, so distant to the first world? That is so wrong.

Anyway, just like the 'Beat the British Drought' campaign I helped to plan for a newspaper back one parched July in the eighties (when the heavens opened and soaked all the papers before the newsagents could get them into their shops), so it was that no sooner had I put the phone down on the Beeb it started to rain: Buckets. The day Barcelona began to ship in water, because the reservoirs had drained so low to reveal lost villages, the heavens opened on the city, buying precious time for the completion of vast underground rainwater tanks and a controversial desalination plant.

Hand on heart, I have made a horrible hash of our irrigation system. It took me three years to find a bung in an overgrown water pipe linked to the spring that, once I'd pulled it free, immediately made redundant the network of pipes I had spent weeks laying in an unsuccessful attempt to get water from the well to our peach trees. I ask you.

Then there was the pomegranate pantomime.

After five and a half years of battling to get water to struggling apple, pear, persimmon and quince trees on the lowest terraces near the road I hit on the idea of transferring, with the aid of gravity, a vast quantity of water from one balsa to the other.

We only have one pomegranate, a single, feeble bush half way between the two balsas that had, up until that point, only ever offered up one fruit on account of unquenched thirst.

So, off I went with the strimmer to clear brambles from around the old pipe joints, to check the route the water torrent would take and to lay new pipes across the corner of the vineyard to the lower terrace. It worked like a dream, and we have had apples and pears, blackberries and pomegranates like never before. My water transfer system leaked more than a British water authority, much to the delight of the pomegranate and the bramble halfway along the route.

The reverse to drought, naturally, is flood. And when it chooses to tip it down it can do so with devastating effect. Welcome? Not always.

If it comes just before the grape harvest and is mixed with stillness and warmth you have the perfect recipe for rot. There have been two particularly terrible harvests, when even some of the perfect bunches had a core of mould. At the end of the hard labour more fruit lay on the ground than in the trailer. At other times the sheer volume and force of water falling on baked ground and running off towards the valley basin means great gullies appear in the middle of our track, and people on remote farms are left high and wet. Mac and Conxita's home is along a two mile track, and through the winter of 2008/9 they were cut off six times, unable to get out, even in their Land Rover

We had more than enough water to last the summer and by late March it seemed things were settling into the old routine of blue or starry skies. Friends from London arrived and we headed off to the Montsant, the holy mountain, drifting through rosemary and broom in search of wild asparagus, lingering by a spring beneath the bulbous cliffs of the range that dominates the Priorat. Back on the farm pine pollen billowed down the valley, bees swarmed and the heralds of heat duelled for attention. Amorous frogs revved like distant motor-cross, and the cicadas purred in the dusk.

Then someone stuck a pin in Zeus again. Flash, crash, bang, wallop. Dentures rattled, trip switches clicked and buckets brimmed yet again.

We scurried to the car to get Joe Joe to the church on Good Friday evening where the village drummers were to follow Jesus on the cross around the village. That was out of the question, of course, so the enormous life-size effigy of The Son Of God, shouldered by one huge guy, did several tight-squeeze circuits inside the church. While thunder roared outside, the drummers thundered within.

It was the hardest downpour of them all. More than twelve inches of rain fell in sixty four hours and our spring turned into a fire hose, spouting 1000 litres an hour.

All of which coincided nicely with the return of Mike (my spring plunge assistant) and Annabel and children Joe and Sophie, which sort of made sense, because something usually happens when they are here. The year before it has been drought, now it was flood. Mike and I went to school together, and long forgotten pranks were hauled from the dark corners of memory, like the round-shouldered failure of the classmate we'd sent into the nearest pub and who was thrown out for asking for "a bottle of drink please".

The sulphur yellow pollen of the pines framed puddles and smudged windscreens. We thought it was some awful toxic disaster until a neighbour explained and I foolishly stood under a branch and vigorously shook it.

Undeterred by the moisture, Mike and I walked across the valley to see Mac and Conxita, a stroll which should take about 20 minutes but which lasted considerably longer on account of his geological training and subsequent curiosity at the mix of granite, sandstone and lime. He pointed to the colossal seam of limestone atop the valley wall.

"That would have been the seabed a few hundred million years ago, of course."

"Uh?"

I have never really been able to get my head around the humongous history of our planet or the obviously relevant galactic and universal inexplicables that equal the meaning of life, despite the Look & Learn magazine subscription my mother renewed until I left home. Geology has never done it for me, although the guy on some TV documentary who explained the periods of Earth's history while stepping from one floor tile to another, finishing with the throat-drying fact that humanity was merely the thickness of the varnish on the final tile, put a lot of tribulations into perspective.

You know, of course, that limestone is an organic, sedimentary rock formed from the remains of tiny shells and micro-skeletons deposited on the sea bed. It says so in books, on blackboards and on the worldwide web. Only here it is towers over everything.

"But look at it, are you sure?" I questioned. "It's colossal. And it's 2500 feet above sea level."

"Yep."

"That would have taken forever." I was sliding into stupor. Mike was sounding more and more like our old geography teacher Harry.

"Hundreds of millions of years. Then, with the changing surface of the planet, it was pushed up into the sky. When people talk of human history, it is but a second in time."

Chin up, his words weighted like at the end of a Shakespearean speech, Mike sauntered on. I stumbled behind gazing over my shoulder as he tapped at granite boulders and mused on how the hillside where we now scratch a living must have once been the bed of a vast river. For me, it was one of those life-the-universe-and-bloody-everything moments, when wondering what it all means just makes your brain hurt and your stomach churn.

Back at the farm I tried to share my feelings with the good woman, but she was distracted by a tree rising from the dead.

Between the late winter downpours there had been one almighty February blow when we battened down the hatches as merciless winds shook all but the rocks. Our valley is a narrow wind tunnel. Part of the wood store roof took wing, pine trees snapped with a crack like gunfire, fences buckled, a diseased walnut succumbed, and Joe Joe witnessed wide-eyed as one particularly venomous gust lifted the prodigious, healthy oak at the end of the lower vineyard out of the ground.

The pony corral began to fall apart too, due to our foolish incorporation of living pine trees. Such was the force tormenting the trees that roots pulsed in the soil and the whole structure swayed and creaked like a galleon in high seas until, inevitably, joints began to fail. To save it, I had to fell at waist-height the most exposed pine that was on the point of crashing down and taking everything with it. Seriously frightening.

But we were spared to say the least. People died. Roofs were ripped off, cars crushed. It rained tiles in the village and across this land. Friends' homes were badly damaged, and I do not know of a lane or road that was not blocked by the swathes of levelled forest. Chainsaws growled across the valley for weeks. As for our beleaguered oak, its demise was dramatic but not lamented. We were more saddened by the death of the walnut just 10 feet from the front door and the loss of the shadow it cast. The cloud of evergreen oak, on the other hand, had blocked our view of the mountain at the end of the valley. And just think of all that firewood, we said.

The headache was that it had fallen across a gulley on to our neighbour's land, flattening several of his hazel trees and blocking access to more. Maybe that was a good thing, because I had no option but to crack on rather than adopting my usual style of dithering until it is too late. If I'd left it any length of time the wood would have turned to iron and been impossible to cut. So for days I laboured, my little chainsaw coping quite well with the still moist branches, until the hazels, pinned to the ground, sprang back, and we had a mighty pile of logs. All but ten feet of the oak trunk was left. I would need reinforcements for that because it was two feet across, with seventy rings. But I had cleared our neighbour's land and I was free, finally, to dither.

Weeks later rain squalls returned and we hid in the house again.

It was Maggie who spotted the oak stump slowly but surely rising back into the air as the water weighed the root clump sufficiently to tip the balance. A few months later new leaves appeared on the one remaining branch.

At the right time, in fair measure, such rains bring bumper harvests, like our first in 2001, and again in 2006 and 2008, when wild flowers cast their beauty and prolific, indefatigable fennel strained six feet into the sky.

After that first year, when we devoted enormous time to the garden, and savoured so much, 2006 was, perhaps, the summer of the greatest unplanned abundance. Late August west winds roared down off the planes of La Mancha and shook the figs from our trees. Empty bags waiting to be filled with harvested almonds were lifted out of the tractor trailer and rolled across the farm like tumble weed as we stumbled about trying to cope with the sudden glut of fruits and nuts.

It is tough enough keeping on top of our own farm – it truly hurts to see hazels lying on the ground unharvested or apples that fall before they can be picked – but when old Juan the fruit farmer calls, we can't resist.

Juan is a puzzle we may never work out; a gentle, private yet kindly man with the greenest of fingers, who grows more fruit than an army would need, yet never sells it, only inviting a few close friends and neighbours to come and share. He just laughs when we tell him to market it. His diabetes means he can longer enjoy the honey from his bees, but on he works. The land, it seems, is his life and the beauty and bounty and the joy of giving are his reward.

So, one Saturday, after we had taken four hundred and seventy kilos of our *garnatxa* grapes to Torres in the morning and harvested almonds for a couple of hours in the afternoon, Maggie somehow found the energy to pick fruit at Juan's, ignoring my suggestion that it was folly to load herself yet further. We had so much of our own to pick and preserve.

Back in the July Juan offered strawberries. The August lure was peaches.

I dropped Maggie off on my way to pick up the children from the riding stables where they spent half that summer, and before we could see the trees we could smell the fruit. All feelings of tiredness and reluctance evaporated. I went with Maggie into the peach orchard in the still twilight and my senses went into orbit. The orchard was so large you could lose yourself. The hues of the changing leaves and giant fruits were heightened in the sanctuary of sunset, while the unsurpassable scent turned the key to lock it all in your mind.

Were that we were so gifted, or organised, or free to think of nothing else for one season at least. Or that I could be so sure of what I was doing as to have mastered such fundamentals as remembering before I watered them that our 350 new little vines near the lane were of a drought-resistant stock. Or that when I had finished imploring them to survive, and apologising at some length for this and a woeful lack of general care, that there hadn't been a man on the lane who had dismounted from his cycle to listen.

The new cabernet vineyard was a folly, for we had little time or spare energy to work it, to care for it sufficiently in its delicate first years. We planted it for both aesthetic and practical reasons on part of what we call the meadow, to complement the holiday cottage, which is called *La Vinya Del Pare* (Father's Vineyard) and to supplement our winemaking. Only about thirty of the four hundred vines have survived due in part to our lack of care, and in part to the fact that three feet beneath the soil is a seam of sandstone.

So the land has returned, naturally, to wild flower meadow which is just as pleasing to the eye and considerably less labour.

A weather warning. Winter in the Iberian sierras can be as chill and draughty as an Eskimo's outhouse.

Serious dollops of snow come on average once every two years, to the point of blocking the track, risking death to olive leaves, and breaking branches. We learned the hard way in 2002. We left snow to freeze on our olive trees and they all had to be cut right back to stumps to regenerate. Maybe some of you saw it on the second No Going Back documentary on Channel Four. We cried. There was no harvest for three years.

Then, in December 2009, the snowfall was so heavy great limbs of pine cracked under the weight. Some olive branches gave way before they could be shaken free, but by wading back and forth with wellies jammed with snow we did manage to limit the havoc. The only problem was we were hit by another massive knee-deep fall in January 2010 that took out power and telephone lines, while frightening winds shook the hilltop village, where normally unnoticed cables strung across narrow streets writhed like snakes, and lamp-posts bucked and twisted. We waited a week for the phone engineers to come, but they were besieged with work, so I walked the line from post to post through our wood, across the wild boar gulley and on to a neighbouring farm, until our link to the outside world coiled to nothing beneath a telltale drift of sawdust beside a stump and trunk. Fortunately those responsible were long gone, (I'm a mild mannered man, but when roused there's hell to pay).

A week later, when the earth finally reappeared, we were almost out of candles and firewood, and still out of contact with the outside world. As the sun shone and the gutters wept buckets, the devastation became clear. The tops of pines that had not snapped off resembled Tom's tail after Jerry had plugged it into the mains. Clouds of brambles were flattened, the ornate ten feet square porch to the cottage was wrecked by four hundred kilos of snow, and everywhere we looked the view had been repainted.

All of Europe suffered, not least the UK, but here it was different in one bizarre way. Instead of widespread stories of blocked drives, black ice, blue fingers and the nation grinding to a halt, all can be radically different just a few kilometres away. While we, at circa one thousand feet, were deluged, the cities of Reus and Tarragona 25 kilometres down the mountain didn't see a flake.

So, there was I, shovel in hand, looking a tad rustic, trying to dig an escape route through the snow by first carving a canyon through the waist-high roadside wall left by the snowplough, when by they glided in slow motion; cardigan sightseers from the tidy suburbs in shiny, climate-controlled saloons. Lots of them. They stared at me open-mouthed from beyond the tinted glass, like they were on some surreal 3D fair ride and I was a Martian.

Imagine it is your average dry or mildly moist day in Dereham or Doncaster but the word is that Wroxham or Worksop has turned into Lapland. You'd be tempted to check it out. But do the afflicted a favour – don't. And certainly don't, as an unbelievable number did here, block tracks and slip lanes to build snowmen and lob snowballs at each other. And as for the gentleman who lent out of a passing government highways 4x4 and laughingly advised me I would be shovelling forever, I hope he appreciated my Churchillian hand signal denoting digging for Victory.

Even if it does not snow, Jack Frost always shows. Our record is minus fifteen. The average nightly freeze through December and January is circa minus seven, but it can linger into February, or return with a vengeance after a mild spell, to steal the almond blossom and our August profit.

If you are ahead of the game and have sufficient wood seasoned and stacked, you can mix bell-clear, frosted days on the land with cosy kitchens and hearty meals cooked for free on the wood-burner. We savour winter moments when we can devote time to outdoors and ourselves, our cheeks ripened by the curious combination of cold and winter sun. Unfortunately, such winter work is often seeking suitable deadwood, for in all these years bar one (thanks to helpers) we have not stockpiled enough to see us through to March or April, when the burners can rest again until November. I didn't used to care, but the enjoyment of scavenging is compromised a little now by other pressures of work, the weave of which has grown ever more intricate as we juggle farm with family, holiday cottage, olive oil exporting and writing.

But the coming of the cold does bring clarity, freedom from insects and moments to mark time. Outside the front of the farmhouse there are six old fig trees, with trunks of great girth and bows that sag to the ground at fruiting time. Apart from weeks when the figs fall faster than we can harvest them, and the flies gather to feast on the sweet fruits that have burst open on impact, we love these trees because they offer summer relief and billow with life. There is also the music of the occasional rain drumming on the broad leaves, the great canopies of which mean the house is all but hidden from view.

One morning in late November I opened an eye, to see the brightening blue through the top window of our high-ceilinged bedroom, then heard this noise, like several people doing a slow handclap with cupped palms. As I got out of bed it seemed to get faster. I went downstairs, unlocked the door, stood back as our hounds stampeded into the light, then followed. The first rays of sun had brimmed the sierra and reached the top leaves of the fig trees, all heavy with ice, and it was the signal for the tree to let them go. A few seconds of sunlight and down they came, almost in a clean, descending line, clattering into other leaves below then piling up on the ground. That was the sound.

There had been no strong winds that year to take the dying leaves, and there was no breeze now to cast them further afield. Everything was perfect during the half an hour of applause that it took to clean the trees.

One dark cloud. We had another late flurry of snow in early March 2010, which closed schools and had me carting my antique seven-foot long skis (last used in 1988) up the land in search of sufficient slope. Joe had been burrowing in the barn for the sledge and had somehow spied the blue Salomon bag stacked against the wall among some old lengths of wood-stove chimney pipe. The old boot case was traced and dusted down, but I hesitated. Something unsavoury had obviously chewed an entrance in one corner. Among the less charming moments of this life was the time Maggie found a plump rat that had passed away in her welly. I put on my thickest work gloves and delved, and there was something nestling in the foam lining of one boot – two very dated ski hats that smelled of...well... the eighties.

The snow barely lasted the day, but there was time enough for first Joe and then Ella to swish down the hill before comically and inconsequentially colliding with tree prunings and broken cane. There was neither time nor inclination to impart the art of turning, and by the end we had worn winter away.

Would that we could so easily erase the ambitions of the mayor of a village about fifteen kilometres from Mother's Garden.

If you look south west on a cloudless day you can see the distant telltale steam plume of a nuclear power station, a pressurized water reactor sited on the bank on the mighty River Ebre. Of the four hundred and thirty six reactors operating in the world eight are in Spain, compared to fifty eight in France, one hundred and four in the US and nineteen in the UK. China currently has just eleven online, but twenty more are being constructed there, while Russia has thirty one with another nine planned. I wonder how clear this picture is for many of us? We all owe to ourselves and our children to dig a little deeper.

Spain's socialist government has been cool on nuclear power and talks gamefully of pursuing alternatives with increased vigour, but the truth here is the same as everywhere where dependency and expectancy are now weaved into the fabric of society. Even if it ended tomorrow there would still be the legacy to deal with.

So, they have asked villages and towns nationwide if any would like to host Spain's nuclear waste dump – blood-drainingly called a *cemeteri* – and the mayor and council in one local community have put their hands up. The village is already home to the power station and so the local council is well aware of the subsidies that would flutter down from the sky. A few other communities across the country have also volunteered, and now it's a case of who can shout the loudest - against.

I let off my steam in a podcast about breakneck nuclear science and a contrary one, that which investigates the principles governing correct or reliable inference, better known as logic. I'm not a political animal but I did this, as well as joining my family, friends, neighbours several thousand other local residents on a protest march, because I wanted to get something off my chest.

Everyone in favour of nuclear power, including the governments that have built the seemingly unstoppable global economic engine that belches toxins and squirrels consequences, makes a fabulous case for it. Read the websites. And do you know what? It does make a whole lot of sense in the here and now, with electrical demand so irreversibly high, the time it will take the trailing sciences of truly "clean" energy to fill the void, and the absence within the nuclear solution of any carbon pollutions or contamination due to their apparent mastering of containment. You want to believe them. Most of the world trusts them.

But (and this is what eats at me) nothing is constant, not least man himself.

Reflect for a while on the calamitous instability of the last one hundred years, and how it would have run if we had been bequeathed such nuclear waste by Roman ingenuity two thousand years ago. Anybody who can predict that neither human madness nor natural disaster will not pose a risk tomorrow or at any time in the millennia, when our successors will have to live with our nuclear legacy, has switched off his or her conscience. That is unforgivable. Our innovative, hungry, wasteful race has littered every room in history and we are now mortgaging the future.

It's hard to put into words, but it is another personal weighty sack of uncertainty about our species, and how, somehow, the voices in all our heads, of reason, moderation, and morality, are drowned out by a collective, induced obsession with more, whatever that is, coming as it does with the hollow consequence of less.

TIME TRAVEL AND PIRATES

Keep your face always toward the sunshine,
and shadows will fall behind you.
                                        Walt Whitman

Ah, España, land of sun and sangria and ex-pats in socks and sandals with a spring in their step. Seafronts clogged with pink or perfectly tanned people from northern Europe trying (and to some enviable degree succeeding) to shed some stress during later life, in the belief that Spain, or any Hawaiian-shirt, happy-hour, deliriously delightful holiday destination for that matter, is synonymous with seamless, twelve months a year serenity. All you have to do is turn a fortnight into a way of life.

Oh, you can, I'm sure, have a good stab at it for a while at least, if you have a small pot of gold, a stout liver, a raft of trusty English-speaking experts to wade through the bureaucracy for you. Find a home on a development devoid of Spaniards and you can dodge the tricky matter of learning the language. But – gulp - imagine what it would be like if the doctor, local lawyer, estate agent and barman hadn't bothered to learn English?

Surreal, don't you think? You can transplant yourself into an English community beside the sparkling Mediterranean and still go out for a Sunday pub lunch - and watch English soaps and the Premier League. You can, as one person said to me, get away from England with all its immigration issues. The irony of that still takes my breath away.

Despite the recession and Euro wobbles, there are still millions from northern Europe here, attempting to live the Costa dream. Fair enough. Positive change is revitalising. Everyone needs sunshine. A great many are making it work, and a fair number are appreciating and engaging with the culture of this country. But, I wonder to what degree they get the chance. Such has been the saturation of English or German or Dutch culture in some hotspots, there is little need or incentive to integrate.

Then there are the likes of us, the rural go-natives, doggy-paddling to keep our noses above water, oscillating between the despairs of misunderstandings and hard labour, and the joys of freedom and fulfilment. We – and there must be thousands of us by now after all the books and new-life television programmes about getting back to the earth somewhere gorgeous – are the ones to ask regarding the romantic dream of owning an olive grove or a little vineyard. There's a fair chance someone living near you in the UK, Holland, Germany etc has been there, done that, because if our area is anything to go by, a great many have returned from whence they came with vivid memories, some light, some as dark as night.

Which begs the question, why return? There are many good reasons, not least emotional and financial, and the simple truth that the minuses have outweighed the plusses.

On the ticklish matter of integration, I think ex-pats fall into four categories. Let's start with the smallest group, those who go totally native, often due to marriage, absorbing and adopting the culture and language and who are as likely to move as George W is to buy a condo in Baghdad. Second, there are those who weave themselves in a considerable way, because they have chosen to try and make a real go of it and because of how easily their children have assimilated, but who all the same remain perched on the fence and could go either way for financial and emotional reasons. Then there are the people of means, living here for a significant time, but not all the time, for whom Spain is either a comfortable retreat or temporary and very palatable place of work. Last are the millions of Costa retirees, some financially comfortable, some wholly reliant on their homeland state pensions and the vagaries of the money markets. Add the misery of planning fraud and bulldozers, and many older generation ex-pats have found that the Latin life can be both as pretty as English porcelain and just as fragile. Even if you have been oh-so careful and weighed up the irresistible charm of a new life against the social, personal and bureaucratic challenges, the butter fingered financial world and unscrupulous property developers can smash everything.

There is another perspective too. Anyone who has been living abroad for any length of time, and has adjusted both their rhythm and attitudes, may be in for some British shocks.

In late October 2006, on the day I was to drive back to England with a friend with a few sacks of almonds, I rose early and tiptoed outside. Clear air glistening with autumn gold, a stillness in which to know how long a minute can be; to do no more than breathe, and look and listen. Indian summers offer such feasts for the senses, though it had gone on so long that year there was talk of fools' gold. Everywhere you could see how the mildness has tricked spring flowers into a deadly flowering.

Twenty-four hours later, I was clinging to the rail of the Boulogne to Dover Speed Ferry, and as the white cliffs hove into view and rain dripped off my nose, I prepared myself for the blast of England.

It duly came, but not just in the climatic sense.

Kent was a picture, as was my sister Jac's old house and garden there, and the trees dripped with colour. After a blast of rain the sun returned too, and my heavy coat was too much to bear. What left me numb was the speed and crush of it all; the density, the mass of cars on minor roads driven by people with white knuckles and wide eyes. It had been many a moon since I'd been anywhere in England save cul-de-sac East Anglia, and I was obviously way out of step.

But that wasn't all. Six years in Spain had made me a liability.

I forgot, you see, the protocol, for want of a better word, regarding contact, or rather non-contact with children. Rather, it never crossed my mind, since the weight of this touchy subject had quadrupled in my absence. The minute I showed up, Jac and I went to greet my niece and nephew out of school. The last of the rain was falling. My niece and her friends came tumbling out of class and she splashed across a patch of lawn into my arms. There was another girl hovering beside her and she was introduced as one of her best friends. I bent my knees to talk to her, shook her hand, began to ask her a few friendly questions, then interrupted myself. The rain was getting heavier and she was carrying her coat over her arm.

"Oh, let's get that on, shall we?" I said rhetorically, and helped her. Then I put my arms round the two girls to guide them back towards the classroom. "Now, let's get out of the rain. I want to see your classroom and where you sit!"

I thought nothing of it. When my sister remarked later that people were stunned and said so, I didn't understand. She said how refreshing it was. She said people were not alarmed since I was clearly a member of the family, but were amazed that anyone could be so uninhibited and spontaneous. That, basically, I didn't think twice about touch. Dear Lord. How utterly sad is that? We talked about the degree to which children are being raised in a culture of no physical contact; about the cause and consequences; about the degree to which fear and suspicion are robbing children of one of the greatest lessons in life, how and when to instinctively use contact in its vital forms of love, comfort or friendship.

I know my children's Catalan teachers, and it is clearly understood that if either Ella or Joe Joe needs comfort it will be given. This world is a very tactile one, both at home and at large, in a Latin culture where contact is second nature. How we give thanks for it. Elsewhere on these pages I spell out all manner to reasons why anyone should think long and hard about such life-change, but the treasures of this existence include seeing Ella and her teenage friends embrace when they meet, and young people doing likewise with family and older friends. Grandparents are visible and enveloped. Communication has not broken down between the generations, yet.

Open arms come with community, of course, in villages and friendly streets, where families stay close and the examples are passed down. Hardly ground-breaking information I am relaying here but, more than Horace's *"piece of land not so very large, where a garden should be and a spring of ever-flowing water near the house"*, it is this fundamental security, sense and balance that I think British people yearn for. It is not about changing countries, but about getting back to values. To acknowledging and being acknowledged.

When I'm invited to talk gibberish on radio here, I usually spout about the social contrast with Britain and how this equally wealth-hungry nation, one that's listening increasingly to financial evangelists about worshipping the god economy whatever the personal cost, has to cherish and protect its fundamental strengths and values.

My trip to England concluded with a few days with my Dad, appropriately, because he possesses one of the finest, most finely-tuned hug-ometers I have ever seen. He reminded me that everyone is fundamentally the same and that you only have to smile at someone to see that.

Nudging ninety years, a widower, registered blind, dependent on two hearing aids and unable to walk more than a few metres without his knees giving up, he lives with the dawn and dusk ritual of pills, eye drops and the puzzle of how he's going to get his socks on and off. But on he goes with open arms and a ready smile, buoyed all year round by the lifejacket of laughter and the sense of touch. He has most assuredly been, I say with pride, a light in his home town of Holt, in the three north Norfolk pensioner clubs he attended and now in the appropriately named Sun Court Nursing Home in Sheringham. Yes, readers of No Going Back might recall, that is where Mum spent her last year. And, yes, it still radiates love, care, warmth and humour.

The hug-ometer I'm talking about is, of course, Dad's serene, beaming face. He needs hugs, which makes him utterly normal. What makes him different and so inspiring is that he openly asks for them. That is why his window to the soul shines like the sun and, hence, does that of everyone willing to drop his or her guard and take him up on his offer. His contentment, and that of other huggers, is wonderfully contagious. So drop your guard and get to it, lovingly, with those you care for, because it will do you and the world a power of good.

Don't believe me? Maybe you were raised to be tactile but are now straight-jacketed by a society increasingly spooked by physical contact of any kind. (One false move and you are in court and, gulp, the newspapers). Or are you borne of handshake parents and already cringing at the mere mention of a hug?   A team from the University of North Carolina in America has scientifically proved that hugging helps both your health and your happiness. A study showed hugs increase levels of oxytocin, a "bonding" hormone, and reduces blood pressure and the risk of heart disease. It also pinpointed an important drop among women in the level of cortisol, a stress hormone.

The British Heart Foundation has said it plainly enough. A hug from a loved one can have beneficial health effects on your ticker.

Dad joined me on my whirlwind tour of shops, and we talked about all this and, finally, the great lateness of the season. He sang, quite beautifully, the ballad *September in the Rain*. All together now...

The leaves are brown, came tumbling down, remember

That September, in the rain...

Almost every May, in the week after the county wine festival three miles away, we have a stall in the ever expanding and entertaining village arts and crafts and wine fair, offering the curious combination of Maggie's homemade elderflower cordial and pizza, cakes by friend Barny, while I perch on the end next to a pile of my books doing very little save gassing with anyone and everyone, as good a place as any to glean all manner of information.

You have to be up before the lark and on parade to help construct cheap and cheerful lightweight canvas pergolas, while somehow trying to guard your pitch in the square or a neighbouring street. Then there is always the pantomime van jam as cheese-makers, glass blowers, barrel makers, wine producers, carpenters, lace makers, blacksmiths etc choose the exact same moment to roll up and unload.

The first year we took part there were no sun shades. We roasted and drank most of the warm cordial. We had to. The locals were understandably wary of foreign faces offering drinks made from wild flowers. The second year it poured, the third there was a saucy westerly blowing that would suddenly gust with alarming force, turning the pergolas into kites. Having known the yachting terror of squalls at sea I kept a weather eye on the distant tree tops for tell-tale waves of wind. It was building not diminishing, so I hared home and prepared some sandbags to tie our pergola down, in the nick of time. There were a few giggles as I puffed up the hill with them, but a few minutes later things were whizzing in all directions and one pergola ended up dangling from the power cables that sag along the street. But still hordes of people drifted by, one or two even buying a book, while beneath the dappled shade of the old plane trees

145

beyond the spring-fed washing pools and safely in the lee of the school wall, seemingly oblivious to the chaos, sat the ranks of lady lace-makers, nattering incessantly while their fingers danced and the delicate lace grew before your eyes.

We made friends with a gentle Colombian selling the hardest cheese in the world, exchanged two cups of cordial for a bottle of wine with vineyard owners Enriqueta and Joan Salvador next door, and took it in turns to drift among the stalls before lunch, the scent or which had been circling for an hour.

A paella dish eight feet in diameter had been bubbling away on the playground, a feast for nearly three hundred people and which was free to stall-holders. In the mingle I found myself beside Maria Delors, whose daughter lives in Mexico City and who has been a friendly face since we turned up. She'd kindly bought my book, and as we queued for our dollop of rice and meat said she'd love to come and visit Mother's Garden some time because she had such happy childhood memories of it. It hadn't been in her family, but she would play with the children of the farm, swimming in the vast balsa and going wild on the land. Come and see us, I urged, and so she did, bringing nuggets of news.

The first revelation made sense. Our house once had four front doors. That much is obvious from the pretty basic building work carried out long before we showed up. "Well," she said. "This lane, although quiet now, used to be the main way up from the coast and your house was a toll house."

"Really?"

"Yes, and you know about the Cyprus trees, of course."

"No."

"Well, it goes back to Roman times. One tree means water is available for the traveller, two means water and bread, and three means a bed too."

Blimey. We have seven. The mind boggles.

"And of course, it is said..." She moved a little closer, "that this house was once owned by a *pirata*."

"What? A pirate?"

She nodded sagely.

"Pull the other one."

"Yes, that is the folklore. They say this was his mountain hideaway."

"But we are thirty steep kilometres from the coast. There's La Mola mountain in the way."

She looked at me, eyebrows in orbit, head tilted, as if I had just made the point for her.

La Mola. It took me five years to get to the nine hundred metre top of it, and I still haven't got to the bottom of it. It is a plateau with vertical white cliffs that, as you trundle along the lane into its shadow, looks like the location for Arthur Conan Doyle's *Lost World* (remember the 1960 epic starring Michael Rennie and Jill St John?), only instead of dinosaurs, lost tribes and verdant jungle, all La Mola boasts is a few hardy pine trees, a shoal of nearly a thousand goats and the mysterious remains of a medieval or older circular stone watch-tower.

The mountain literally dominates our life here. We are all in the habit of stopping and staring at it – from the house, from the top of our farm, from other high points of the rugged hinterland, from the windows of jet planes as they climb away from Reus Airport, and from the distant towns and beaches of the Costa Dorada.

It stands so tall and proud that when I see it from the motorway at Tarragona, forty minutes still to go until I reach Mother's Garden, I feel I'm almost home.

I declared soon after we showed up that I would conquer it, but for an embarrassingly long time never got close. Something the height of Scafell Pike in Cumbria requires a run at it, a little bit of focus and time and, naturally, suitable weather conditions. The climb is, er, steep, and a baking midsummer assault would have been as stupid as attempting it on a bleak day. In winter the plateau is often cloaked by cloud, topped with snow, or swept by icy winds.

Very dodgy. That left the spring and autumn, the months of hard farm labour, so for years I was all talk and trousers.

Pathetic really. It was as unlikely as someone from Blaenau Ffestiniog never bothering to do a Tarzan call from the summit on Snowdonia, or Boris Johnson being bored by the idea of a ride on the London Eye. And, in case you were wondering, Jeremy, it had nothing to do with nose-bleeds and being raised in Norfolk, which isn't flat I'll have you know. Flattish maybe, but definitely not flat. Truth be known, I was born in the lee of the county's highest point. Oh yes. At three hundred and thirty eight feet Beacon Hill at West Runton is a veritable peak, with such commanding views of the coast that it was a site for a signal station during the Napoleonic Wars.

It took Maggie's brother Phil, a fit-as-a-flea front-row member of the Mother's Garden life-support team, who has criss-crossed the British Isles on Shanks's pony and who devotes most of his supposed leisurely sojourns with us strimming and lugging timber, to make it happen. He's also once brought a band of fellow high-mileage members of the Norfolk Hill Walking Club to see us. Yes, the Norfolk hill walkers. Unlike the Gobi Desert Canoe Club they do exist. They are a happy band of 30-plus people who have stepped the great trails of Britain and beyond, who breathe the best air and know the health truths and fulfilment of being out there in the heart-pumping real world. High-mileage is no euphemism for age. It is a pedestrian reference to experience and rosy cheeks. Andrew, Lynda, Richard, Ian and Phil strode out on the ancient trails, meeting no-one, returning replete with smiles, bird sightings and the warning of one cliff-face route that would make an ibex hesitate. It was so precipitous you had to hang on to a cable bolted to the rock. They reminded us to get out there, they taught us where to go and they inspired us to make far more of the "walking holiday" appeal of Mother's Garden. It all went as smoothly as north Norfolk Morston marsh mud between the toes. And at the end of the week, at their behest, I took them on a birding outing to the Ebre Delta where, as on the north

Norfolk creeks and pools of my youth (but for the seam of flamingo pink), the scent of salt and serenity stopped the clock.

With a serious assist from Maggie who looked me in the eye and told me to stop faffing about, I dug out my hefty clodhopper work boots that were the nearest I had to hiking gear, and loaded a rucksack with water bottles and mandarins. The good news was that we could drive halfway up La Mola, to the fifty wind turbines that line the ridge running north from the mountain. They hummed in the breeze as I locked the car door and looked up. It didn't seem like we were any nearer, and now that we were up close, the climb looked alarmingly vertical.

On the level I'm reasonably fit. The only trouble was we weren't, and after the first two hundred metres of steep rocky path through wild rosemary and thyme I gave up glancing at the view and concentrated hard on keeping up and not muttering. My steel-capped boots, rather than trainers, were a serious error. We zigzagged through a spooky wood of stunted oaks passed stone ruins and boar tracks, then it was a nauseatingly slow feet and hands climb across a rock face where a former goat that had lost its footing pulsed with maggots and flies. Beyond the stench I tried to get Phil to stop on an outcrop, for pictures, to admire the view, drink water, and gasp enough air to diminish the pains in my chest and limbs, but the dear boy had the bit between his teeth. The last section was the worst in every sense barring dead animals – a path of sorts through steep loose rocks and shale up through a cleft in the cliff face, complete with a dodgy moment when I lost my footing.

But, boy was it worth every step, every gasp. Just over the lip, right in front of us, was the ancient, possibly Roman, certainly medieval, fortified watchtower. Whoever's misfortune it was to man it must have been faced with the colossal problems of supplies, staving off hypothermia etc, but what a commanding view of the watery horizon, the shoreline and fertile coastal lands that have been fought over for millennia. The Roman city and thriving port of Tarragona, lay beneath us, along with the creeping sprawl of Salou and other smaller resorts and ports. Any war fleet with ambitions on Tarragona would have been spotted long before they reached the beach, unless, of course, they arrived on a cloudy day, in which case the sentries on La Mola would have been having a truly miserable day at the office.

Our day, though, was crystal clear and eerily quiet, and after a moment out of the wind by the tower we decided to yomp to the far end of the plateau to try and spy the farm. It was a fifteen minute fast walk, made all the easier because it was across bouncy Sound of Music grass. A pair of ravens soaring above us, riding the thermal from the cliff, but no eagles.

I've returned, with the children and friends, but it has never been as peaceful and magical as that first time. Once, with Maggie and a group numbering several children, we had to turn back at what has become known as the goat rocks because cloud closed in. In the blink of an eye the temperature plunged, the rocks became slippery and we couldn't see further than a few metres.

You can imagine, then, how long it took us to tackle the Pyrenees. Nearly nine years.

They are only three hours north of us, those dreamy heralds of holiday, the ribbon of snow caps or jagged summer peaks that draw wide eyes to jet windows. Hands up all those who haven't stared down on the Pyrenees in vague bewilderment at our balloon of a world? An easier number to count, I wager, than those who have.

These have to be among the most wondered mountains in the world. Every week tens of thousands of souls lose all sense of proportion as in a matter of minutes they breeze over one of nature's great barriers before commencing their descent to sangria Shangri-La. At five hundred miles an hour and from six miles in the sky the ragged line that separates verdant France from Iberia seems so brief. Mind you, what is real when we are up where we were never designed to be?

We tackled the Pyrenees at about thirty miles an hour, then parked in the village of Boi, squeezed into a 4x4 taxi to bounce ever upwards into the national park known as twisted waters (*Aigüestortes*), wandered a while then flopped to a happy halt for a picnic on cushioned grass beside a brook. In the twenty-four-thousand acre park that's walled by mountains in the Val D'Aran, west of Andorra and almost abutting Aragon and France, there are some two hundred and seventy two lakes and countless pools, cascades and wandering streams left by glaciers. It is a Catalan ecological treasure trove which, if kissed by sunshine, has fairytale beauty. Dun cows with bells the size of loaves laze as a trickle of people trek past in rolled socks and sensible shoes, either for a day hike or en route to one of the ten park refuges. Vultures – I saw Griffins but sadly no Lammergeiers – swirl around the 10,000ft summits. Forests of pine, fir, birch and beech are stubble on the cheeks of sleeping giants whose noses breathe in the sky. It has the majesty of all great altitudes, and is the home of a mosaic of plants, insects and mammals, among them the goat-antelope Chamois, the ground squirrel Marmot, and huge and hairy (sic) Brown Bear.

If you have never been, I urge you to go. Our experience reinforced the common feeling, that time is short and there is so much to see.

Every September, after three months of back-to-back cottage visitors, in the last gasp of the children's oh-so-long holiday, after the almond and before the grape and olive harvests, we crave to get away. That currently means squaring the common family circle of Maggie's and my increasing need to sleep late, read, lunch, snooze, stroll, dine, read and sleep, with Ella and Joe Joe's youthful thirst for experience and excitement. This feat is always further complicated by our flat finances. The Pyrenees might just have worked if we'd opted for campsite bungalows somewhere, but we were neither sure where to stay nor what to target. I don't know about you but we hate being herded, penned and paying through the nose for something unfeeling and clearly tailored for tourists. We were all set to take pot luck when friend Marivi popped the question. Would we like to be her and husband Carlos's guests?

Marivi is our children's doctor, the one who, back in 2001, roared with laughter at our expressions when we required six-year-old Ella to translate for us. Somehow, in an inordinately brief time, our little girl knew enough Catalan to explain that Marivi needed "to put a bag on his willy and collect come peepee." Marivi has also been Ella's English student, and during one of their weekly pronunciation and grammar workouts at the garden table Ella mentioned our excursion.

"But we will be there too!" said Marivi. "We have two little flats in a village. Stay with us."

Bingo. And just in case both our spring-loaded offspring dallied with precipitous danger, in unison it was doubly comforting to know Carlos was a paediatrician too. So, from vague wishful thinking to certainty in one happy moment, and for three cloudless days Marivi and Carlos guided us to those places you rarely find without the compass of local knowledge. We visited one dizzily high village where, in a farmhouse museum that was home to twelve generations of a hardy mountain dynasty, the elders were afforded the warmest bed - in a kitchen cupboard beneath the stairs. They guided us to a traditional Pyrenean restaurant with scrumptious set menu that we would never have discovered in a month of Sundays, and then to that brook in the high valley of twisted waters.

A bow-legged farmer in their village of bowing slate roofs stopped to talk. I lost track in the byways of dialect, just as a foreigner might on Swaffham, Selby, Barnstaple or Brixton markets. It turned out he was recounting the impoverished, dark decades after the civil war, when Franco fenced off the nation and punished his opposers, not least the

Catalans. Life closed in, more so in the tiny, isolated mountain communities. People could not afford motor transport, so the gentle man of the southern slopes had made a living breeding and trading in mules, just a few miles from a radically different world called France.

On the last day we fulfilled our equestrian promise to Joe Joe and Ella and saddled up. I hauled myself up onto the back of a steed for first time in 25 years and off we went, me looking ludicrous in a cycle helmet because nothing else would fit, weaving between lines of poplar trees and crossing and re-crossing rivers along the narrow and lush valley base at Vilaller.

Marivi was born in León in north western Spain. Her father, also a doctor, moved the family to Barcelona when their hometown was drowned by one of Franco's punitive reservoir projects. Carlos is Basque. They met when studying, and have been caring for children in the communities close to Mother's Garden all their careers. One evening, while Carlos, Joe Joe and Ella fished fruitlessly in a mirror lake, Marivi, Maggie and I walked her dog beside a shallow stream along the frill of the forest looking for – and gleefully finding - wild mushrooms. The murmuring water had the delicate hue of the blue green lichen jewelling the bark of dead branches. In one glade we counted ten different species of trees.

In the fading light the forest brooded. I mused, half-jokingly in a "reassure me" manner, how bears might be watching us. Fat chance, of course, given there are only about twenty between the Atlantic and Mediterranean.

"A mother and cubs appeared quite recently, actually," Marivi replied without a care in the world.

"Where?"

"Very near here."

Come to think of it, we haven't had more than a week away since the children were born. Visits to Britain don't count. More often than not they are a panicky patchwork of family, friends and work (olive oil deliveries or meetings with publishers), haring hither and thither, when we try to do too much and always return torn and with a sense of failure at some level, usually to do with nearest and dearest. Lack of a long leisure break is partly financial, but equally reflects the life we have weaved, featuring a menagerie it would be unreasonable to ask others to cope with for any length of time, and a bio-diverse farm that has Forth Bridge attributes. Anyway, I don't like being away too long. I'd miss something.

On the matter of bio-diversity, though, I'd beg for the money and somehow find the time to visit the United Nations Assembly in Manhattan, if they would but let me address them. I'd be brief and to the point.

I have just re-watched the film "1984". It is twenty years since I last saw it, thirty or more since I read the book. It still sends a shiver down my spine. How I wish Eric Arthur Blair was alive. He would take up the pen to wage polemic journalism and, I dream, produce a novel to shake Spain, Europe and the World from their stupor. You and I know him by his literary name, George Orwell, author of Homage to Catalonia, The Road to Wigan Pier, Animal Farm. His renowned loathing of social injustice and totalitarianism, together with the great weight of his words, are so desperately needed now.

The issue is the oligarchy of agricultural "innovation", and with it the brilliantly insane, money-making genetic manipulation of the natural world.

I have made clear my abhorrence of nuclear power. It is extraordinary ingenuity, but we have no conclusive, clean answer to the legacies. Genetic modification of crops is its food chain parallel. Neither of these world-changing sciences stands the simple test of sense.

You knew (didn't you?) that Spain is by a very, VERY long way Europe's largest producer of GM harvests, a staggering 80,000 hectares at the last count. Here, production is ten times higher than any other country on the continent.

The corn fields of Aragon, where Orwell fought for the Republic, is now the battleground for awareness regarding the widespread cultivation of Bt maize. Bt, *bacillus thuringiensis,* is a soil bacteria and scientists have incorporated a gene from it into the plants that now give out a toxin that is poisonous to pests.

That is just one seed of plant "redesign" among many. You are no doubt fully aware which products you buy contain this GM corn. But, in truth, we common folk don't know, do we? And not even the few dominant, emphatically profit-driven giant corporations who have opened this box of irreversible mysteries can know the long-term consequences.

I am writing this on June 5, 2010. World Environment Day has been on the fifth of June for years, but did it register with you?

Biodiversity – safeguarding the variety of life on Earth - is so, so, so important, and 2010 is the International Year of Biodiversity. But how many people know what the word means and, indeed, care what bearing GM science could have on it?

Oh, how I wish you would. I bet many of you who haven't explored the topic will be stunned, engrossed and motivated. It is not a 1984 thought crime to search for more information, to educate ourselves about our food, our besieged environment. Big Brother is not watching you. Think long and hard about what you are putting in your body and ask yourself simply, what are we allowing to happen? History surely teaches us to respect nature and to work with it in balance and harmony. But we are far too clever a species to learn. The greed is for profit. It is, for sure, the stuff of an Orwellian novel.

I'm relatively optimistic by nature, but the glaringly obvious Planet Earth problems of mankind's making depress me so much that I regularly have to walk the land counting the types of grasses and considering wild blossoms, among them camomile, calendula and corn flowers from the gardens of Roman herbalists.

"Biodiversity is life. Biodiversity is our life" - that is the United Nation's accurate but, I fear, forlorn attempt to wrestle the unseeing, desensitised, materialistically pre-occupied world out of its bubble of indifference. I say forlorn because this "slogan" for the International Year of Biodiversity remains faint when it should be shockingly LOUD, a cry rising to a clamour. I used to think that it required something awful to happen for there to be fundamental awareness and change. Yet, awful things are happening all around us, because of us. The difference is that the decay still doesn't directly affect the critical mass; yet. Due to overriding economic structures and social pressures and conveniences it neither alarms nor inspires radical thought. So, I've just walked the land to blow out my cheeks and try and dwell in possibility, before re-reading what exactly UN Secretary-General Ban Ki-moon had to say on survival. It helped.

"Our lives depend on biological diversity. Species and eco systems are disappearing at an unsustainable rate. We humans are the cause. The consequences will be profound."

It should be on billboards, television screens, T-shirts and school walls. But first of all it needs to be understood.

mother's garden

ANOTHER COUNTRY

Homesickness is... absolutely nothing. Fifty
percent of the people in the world are homesick
all the time... You don't really long for another
country. You long for something in yourself that
you don't have, or haven't been able to find.

John Cheever

I savour occasional sojourns of anonymity down the mountain
in the maze of Reus, our nearest city. Its modest heart of alleys
and squares is a place of gentle beauty with no thumping
landmarks so, subsequently, has nothing like the tourist hordes
of its neighbour Tarragona five miles away.

The charmed time is the awakening hour, for the city
centre opens an eye and stretches with such beguiling easiness
it feels like Sunday morning in the heart of York or Norwich,
and is rather homely. If the car has been burping, farting or
flashing a warning light to signify we must give another wad of
dosh to a garage, I have nearly two hours between the garage
opening and the train departing. The last time this was
necessary, I made my way as fast as possible through the
frantic business fringe of the city, then stopped on the
pavement outside a petrol station to admire a leaf-green 1978
Mercedes 300 saloon with tan interior that looked like it had
been moth-balled for three decades. (I used to have an old

Mercedes, you see; called Henry). A man of enormous girth, contented jowls and car seat complexion sauntered up to it and then glided away, heading into town. I followed, and round a couple of corners there was the car, abandoned in the road, inaudibly ticking over. The driver's door was open, but there was no sign of Mr Hitchcock.

I veered into a newsagents/tobacconists to buy a sports newspaper and there he was, lovingly scenting six suitably fat Cuban cigars before delicately loading five of them into the shafts of a silver case, then lighting up and indulging in the sixth like a Nineteenth Century earl in a gentlemen's club. The shop filled, but nobody muttered.

A little further in and the city grew quieter still, with the trappings of later crush – crowds of bar chairs and tables – all but deserted, what coffees there were being stirred oh so slowly as eyes drifted through morning papers. The offices of a political party were patterned quite artistically by red paint bombs. Men in yellow overalls were moving steadily through the streets, sedately sweeping and emptying bins. I stopped at my favourite watering hole and glanced from my Barcelona football newspaper to one of many signs above shop doors that pulse the time and temperature – twenty one degrees Centigrade at 8.45am, twenty five degrees by 9.30am. Outside the air was heavier, the smells more stringent. The feeling of comfort slipped, and I took the train back into the hills.

American novelist Cheever's words are thought-provoking, don't you think? I firmly believe home to be as he defines it. The secret is resolution, which so often resides just around the next bend.

We came here, to an altogether different country, to set up home. We have put our hearts and souls into it. We have pushed through all manner of nettle beds, knocked holes in walls, bled, cried, weeded vineyards by hand, filled countless albums with priceless memories, forged indelible friendships, touched the land, seen our children benefit in all the ways we had hoped, yet we are still struggling with the home equation.

It helps to reflect on where I have hung my hat over the years. The bricks and mortar we left in England meant so much to me because I had vital memories spanning twelve years of really growing up through my late twenties and all of my thirties; of isolation, learning to live with myself, restoring the cottage, Maggie arriving and our lives being enveloped in the brimful emotions of parenthood. I truly thought I'd have to avoid going within a mile of it, fearing raw emotion. Yet I have had to drive past it occasionally when we have visited our lovely former neighbours, Arthur and Audrey, and it has stirred absolutely nothing. I and my hat have moved on. Home for me is about people. And at this nucleus time of upbringing that means, whole-heartedly, Maggie, Ella and Joe Joe.

All the same, it is never a good day when the best bits of an English existence are nudged to the forefront of our minds by some unpalatable but none the less pertinent truths about going native in remote Iberia – the thumping July and August heat, day and night; flies in your face and on your food; dogs across the valley howling night after night; arms, legs and faces swelling after mosquito bites, legendary broad Spanish red tape; or the occasional communication meltdown when the language fails you when you most need it.

Our antidote to such anxieties (invariably borne of over-tiredness and overload) is a fix of words or music or both, from reading in bed on a Sunday morning, to taking ourselves off to the *Palau De La Musica* and other venues in Barcelona or nearer to home for concerts. In addition to its attributes of beauty and peace, Mother's Garden is close enough to the hum of civilisation for us not to be starved of culture. We just don't reach out for it often enough.

The freedom of self-reliance can be intoxicating at first, but divorcing yourself completely from the regimented world is some challenge, as any self-employed person will tell you. Being disciplined regarding work is important (so I have been told), but being disciplined about non-work is absolutely crucial. We do this in fits and starts, and after a decade are starting to get the hang of it.

Holidays, as I have said, are a prime example. The lesson is that any break, however brief, always reinforces the glorious things about home.

Imagine our delight, then, when we were told we had won a holiday. We were that knackered, broke and in need of a breather that we believed it almost immediately.

I suppose if you were to draw up a list of things that are never quite what they seem, winning a holiday would be up there along with election promises, wooden fruit and The Surprising Adventures of Baron Munchausen. Hands up who has been told they have won when they haven't even entered? My mother-in-law receives such good news in the post most weeks. So, when the pleasant Manel who had supplied the Swiss Ruegg woodstove for the holiday cottage rang me on a particularly knee-buckling hot August day to ask if we were happy with the woodstove and to congratulate us on our luck, I naturally thought it was a wind-up. I wiped the sweat from my brow and mustered the Catalan for "you must be having a laugh", but it proved completely cast iron. Our names had been plucked out of the proverbial hat for a family holiday on the Riviera Maya, Mexico.

We hesitated, even so, for four reasonable reasons. It was a ludicrously long way to go just for just a week (in January); it would involve a long-haul flight; we had to get ourselves to Madrid on the train and cough up for an airport hotel, all of which totalled several hundred Euros; and there was our reluctance to pump Ella and Joe Joe with vaccines. But, go we did (without vaccines) in the name of gift horses and the lure of the wider world, albeit packaged. One minute I was in wellies, Alaskan hat and winter thermals feeding the ponies, the next I had nothing on save Eric Morecambe shorts, standing ankle deep in the crystal clear Caribbean, continuing the family tradition of beachcombing meditation. Back in the polka dot bikini sixties my mum's Sheringham guesthouse featured countless, priceless shell-decorated bedside lamps that - guess what - had once been Tia Maria bottles.

She would have loved it. Very posh it was, oh yes, with beachcombing the likes I doubt I will ever savour again.

Our Air Europa flight from Madrid landed at the same time as a squadron of Canadian charters, and before we'd left the airport we had a pretty shrewd idea of the dominant culture. It wasn't Mexican. It seemed the Riviera Maya is to the Canadians what the Costas are to the Brits. As we waited for our bags to appear five-year-old Joe Joe had wandered over to another conveyor belt where a kaleidoscope of countless gaudy vanity cases were coming through the plastic curtain. "Where are all the little people?" he asked.

On our last day we stayed on the beach and drifted away from the Jamaican rum bar, pools and plastic sunbeds, drawing breath for the twenty-two-hour coach-plane-train-car trek home the next day, reconciled that our bumpy coach trip and short trek through the jungle to the lost Mayan city of Coba, along with snorkelling through great shoals of fish and watching Maggie and Ella swimming with dolphins, added up to justification for such a holiday.

Actually, we were also trying to put some distance between ourselves and some whooping Canadians who clearly relished the highly luxurious, non-stop food, fizz and five-star beach resort frivolity that had among the imperatives of fun incessant American football coverage on the big screen TV.

'After your exercise class, would madam care for another pina colada and have her photograph taken with this iguana on her shoulder?'

Not that we were ungrateful.

No doubt, in this shrinking world, some of you are familiar with Mexico's Yucatan Peninsular that reaches towards Cuba and forms the north-west shore of the Caribbean. Once a great coral reef, and later a Mayan kingdom plundered by the Spanish conquistadors, this limestone flatland of mangrove forest and white sand beaches has become a tourism hotspot with Cancun as the hub.

Just before the Kirbillies descended, the area was on the world news because Hurricane Wilma had thwacked it between the eyes, leaving several luxury hotels shattered. We, somehow, were booked to stay an hour away down the coast towards Belize, at a place called Akumal, where storm damage was limited to palm trees and overhead wires. Akumal means turtle beach, and is where the gentle giants of the sea labour up the sand every summer to lay their eggs. Hand on heart, the resort was surreal, and a tad uncomfortable at times for we simple folk. The best moments at the hotel were watching the pair of ospreys fishing on the reef, seeing squadrons of pelicans glide overhead, listening to the birdsong, sailing a catamaran a mile out to sea, meeting some gorgeous Mexicans, watching the children's glee at all the things to do, and walking the shore. You should have seen, though, how the smiles of the extremely tidy Canadians in the next room fractured when we spilled the beans that we were phonies who hadn't paid full whack for the privilege of having someone fold and shape our bath towels into the form of mermaids, elephants and swans. The mother was a shoulders back, knees up, arm pumping pre-breakfast jogger, her subtle tan offset perfectly by pure white sneakers, socks, shorts t-shirt and head band. She would, bless her, accelerate or u-turn if she glimpsed un-ironed, stubbly me wandering back from the daybreak buffet bar with two cups of tea.

Along the lip of ocean Maggie and I wandered on the last full day, seeking nature's sound and admiring but not, I stress, taking the broken coral that Wilma had thrown up into the shallows. Two large senior German women in swimsuits and armed with massive colanders had different ideas, and were working the rocks with alarming efficiency.

Then there it was, between my feet in a small circle of sand in the dead coral, in about six inches of water. In the relative calm between the waves a small stone sticking slightly proud of the sand caught my eye because it had two perfectly perpendicular lines on it, a strange order among the debris. I dug my hand into the sand and pulled it out. The stone was covered in grains and I dipped it back into the water to wash it. Then it slowly dawned on me what I was holding. In my palm was an ancient Mayan carving of a head, possibly a king, as I later learned was the Indians' custom. It had survived the centuries and the storms well enough for me to discern the nose, eyes, and hair, and that it had been polished. I mouthed to Maggie that I was about to have a cardiac arrest and she paddled across to see. We then sat on the sand wondering what to do, me staring at the carving and running my fingers over it. First I drew a map of exactly where I'd found it, then we went in search of someone with archaeological knowledge to look after it and, hopefully, tell us something about it.

Finding them proved a lot harder than finding the artefact. Having spoken to six people we were growing very anxious that nobody seemed to be taking the find very seriously. We were even advised to take it home as a souvenir, which we would never do.

So, with the carving watching over us from the table at the end of the bed that night, we decided we would check out early, let Ella and Joe loose in the kids' club for a while, and attempt to shoot off on a local bus to a museum further down the coast; all this on the day we were due to start an eight thousand kilometre journey in the opposite direction. There was no telling if we would get back in time to bundle up the kids and catch our coach-link to the airport, but there didn't appear to be a choice.

Maggie is one who laughs in the face of such dalliances with lateness in such a way I think she enjoys it. She once had me taking a train jaunt out of Dublin and down the east coast of Ireland on the day we were due to fly back to England. I, on the other hand am only happy if I'm sitting on my case at the check-in an hour before it opens. Thank the Mayan gods, then, for Lupita. She was at the cashiers' desk when we handed in our keys and paid up what we owed, and she asked why we were checking out so early. We showed her the stone, and she immediately said she knew that the director of the diving centre along the beach, one tall, handsome Octavio Del Rio, was a marine archaeologist.

It was Sunday morning, but he was at work. He came. He saw. He went into orbit. And Maggie swooned. He took us back to his office, showed us the article in National Geographic about his underwater discoveries, explained how he would alert the head of the government's sub-aqua archaeological unit, and we sighed with relief, dreaming of the day we might return to take him up on his offer to dive with him off the reef.

Mother's Garden dusting duties aren't helped by our reading habits. There is a bookcase in the kitchen, one in the hall, two in the office, one halfway up the stairs, half a wall of books in the living room, shelves in each of the children's rooms and two bookcases in our bedroom. Oh, and about six-boxfuls in the loft, plus, needless to say, haphazard stacks by the beds (besides Sunday newspaper supplements) and in the bathroom that are never constant from one day to the next. They are the sum of contentment, a storehouse of wonders, fictional and factual that grows indiscernibly; a myriad of little doors to other worlds, a pattern of colours and vertical words that for me surpasses all art. Time evaporates when I am reading or when I walk through the door of a second-hand bookshop.

When we read to each other on a Sunday morning, it is invariably natural history tomes from the early and mid 20th century. We appreciate all fine prose, but my literary heroes are headed by William Dutt, Lilias Rider Haggard, Donald Culross Peattie, Adrian Bell and Ted Ellis, who walk at my pace in the sort of places I need.

Only Donald is different. I mentioned him in the *Pickles* chapter. Look him up on the internet. Better still, try and source one of his books about north America. I'm familiar all my life with the east of England land and waterscapes of Dutt, Rider Haggard, Bell and Ellis, but we had to go in search of Donald, in Provence. It was in vain. He wrote a history book about Vence where he scraped a living in the 1920s, but nobody in the tourist office or library had heard of him. So we pottered about for a day imagining him there, and bought a bottle opener and cheese cutter from a junk stall to trigger thoughts every time we indulge.

It's always slightly strange being in France. For one thing, I have completely lost any capacity to speak the language; for another, our first plan was to live there, in the Garonne region, not in Catalonia. How weird to imagine that, but for fate, the children would be arguing and carving a life in that language. Readers of *No Going Back – Journey to Mother's Garden* will remember how I had found a little wreck of a French farm for Maggie to see, and that we had only spun south into deepest Catalonia to meet up with her sister who was staying with Mac and Conxita. They told us there was a farm for sale close by. Ho ho, we laughed. Seriously, they said. We saw Mother's Garden and that was it. France evaporated.

And in ten years I have only glimpsed that vast nation three times, Maggie twice. As well as our Provence holiday the other occasions have been break-neck drive-backs to England.

This is the tale of our Maggie and my first road journey between here and the old country since the Kirbillies convoy of one knackered Range Rover with roof-box and trailer, a seven tonne truck and a friend's van rolled up at Mother's Garden in January 2001.

It seemed like a good idea at the time, but despite packing the Thermos and sandwiches I can't say it was a picnic. We had to justify it, of course, carbon footprint and all, and that is where the wheel almost came off. Finding good reason was simple; to promote our olive oil business and to lug a few bags of our untreated almonds, walnuts and hazels. But if it was to be for business not pleasure then it would have to be just the two of us, because we would have to stack the rest of the car to the gunnels with new harvest olive oil for food writers, chefs and key customers.

Somehow, Lord knows how, it was reconciled that Ella and Joe Joe would stay at home in the care of trusted friends and help look after the animals, while pressing on at school. We would return home in the shortest possible time, nine days later, with a car still weighed down, only this time with my great-grandfather's chair and assorted other important possessions that we couldn't wedge into the truck way back in 2001, when we headed for the hills. Leaving the children at home sort of makes sense, doesn't it? That's what we kept telling ourselves. And to make absolutely sure there was no niggling doubt, we nailed down a tight schedule worthy of a boot camp. You may have a shrewd idea where this is heading.

10.00pm Monday. We were going to leave at midnight, but notoriously cautious me brought the time forward. I'd spent the day treble checking that our Opel Zafira wasn't over weight before I grabbed an afternoon snooze. Maggie had drafted a day-by-day task list for everyone staying on the farm and we were set. We hugged the children, tucked them into bed saying that by the time they got up we'd be half way there, trundled into the night and were immediately gripped with a horrible gnawing guilt and doubt. It all suddenly seemed bonkers and frightening. We tried to reassure each other that all would be well, Maggie dozed off, I drove as far as the French border and then Ella sent a text saying she was missing us already.

6.00am Tuesday. All of you hardened European drivers will know that a) France is vast, and b) the relatively new A75 motorway across the Massif Central between Montpellier and Clermont Ferrand is quiet and toll free but for the v-v-vertigo Millau Bridge. It's also a road with altitude, the clue being in the word "massif". For circa two hundred and fifty kilometres we only saw a handful of other vehicles, mostly trucks wheezing up hills. Fab, I smiled. Open road, free motoring, and ahead of schedule. At the highest point, 1121 metres, 3677ft, the car lights began to bounce back from snow on the verges, but the road was dry and we were beginning to drop down towards Clermont Ferrand .

Then just before the town of Saint Flour – wallop – white out.

7.00am. We crawled into a service area, sipped coffee with an assortment of other miserable souls, and watched as the day struggled to break. There were fluffy flakes of Bing Crosby proportions and icicles hanging from car bumpers. Maggie very kindly didn't ask why I had failed to figure out that the cheapest route might, in mid winter, be the stupidest.

That's it, I thought. Our non-stop whizz through France, and consequently the whole trip, were over before they had barely started. I whiled away the time tapping and fiddling with the sat nav. It hadn't mentioned snow. I'd bought it especially for the journey to limit stress. The woman inside it who told me where to turn, spoke very confidently. She never mentioned snow.

8.00am. "Maybe it will clear," said Maggie. "It might not be so bad further on. Let's try."

I mumbled and grumbled back to the Opel, we returned gingerly to the motorway and tickled along at thirty mph. Calais was still a huge dollop of France away and I'd booked the ferry for 6.30pm.

8.30am. Hail she who must be obeyed! The weather and roads cleared and the car and the driver rapidly got a grip of themselves. The next conundrum was whether to navigate cross-country via Chartres to Rouen and up the coast, or to take on Paris. Neither looked easy given that I'd dropped the sat nav which now wouldn't work, and our map of France dated from well into the last century.

10.00am. Another short stop, and before dropping the sat nav into a bin, I noticed a card thingy poking out of it and tried pushing it in. The woman within it immediately woke up and told me to bear left and then turn right.

3.40pm. Calais. I'd have kissed the sat nav woman if she'd materialised. Paris was a piece of cake. We could board an earlier P&O ferry. The rain thrashed down, but we didn't care. It was blowing a force seven but we were oblivious. By 7.00pm, UK time, we would be at my sister's home near Canterbury for a hearty supper and then bed.

Wednesday. Brighton. Traffic horrendous, but happy hugs with our niece and other friends, before heading west and then north through Shoreham, past the Hogwarts-esque

Lancing College, into Upper Beeding – "Hang on a minute, that rings a faint bell".

My godfather, once a master at Gresham's School in Holt, and later at Lancing, who I haven't seen for forty five years, sends me birthday and Christmas cards from his home thereabouts. We had no time to track him down, sadly, but I was mildly chuffed all the same that my brain was still that alert. On the Sussex/Surrey border we met with a top food writer, ate scones in her conservatory as night fell in the late afternoon, then attempted to high-tail it to one of our top oil customers in Hampshire. Rush hour bedlam. Our progress that black evening and the following damp day into London was both frustrating and alarming.

Thursday. London congestion charge £8, hourly parking £4. We dropped oil at my old publishers, and then for opera singers at Covent Garden, called at a customer in Crouch End, before resetting Miss Satnav to get us to Finchley where, unbelievably, we somehow squeezed a drum kit into the car – summer guests to the holiday house had said we could have it for Joe Joe. If that wasn't enough to crease us, we tried and failed miserably to make it out of log-jammed Enfield to a 5.30pm appointment in Epping High Street before pressing on to Norfolk for the night. Ha, B***** Ha. Twitch, twitch.

Friday. This was supposed to be a quiet day. We headed into Norwich where I had a coffee and chat with creative director Frank Prendergast at Eye Film to talk about the film project, before midday hunger was sated magnificently, it has to be said, by fish and chips on the market. We then delivered oil for Delia and head chef Chris Pope at Canary Catering, before Maggie spotted a closing down Woolworths store, found a trolley-load of bargains (gifts, kettle etc) then zoomed off to another shop leaving me to queue and pay just as the checkout computer rolled over.

We went on to plunder first a health and then an electrical store and filled the car with vast quantities of ecological washing powder and toilet cleaner, organic bread flour, oats and a bagless vacuum cleaner.

Saturday. My old cub reporter turf of Swaffham. First to the glorious Green Parrot Health Food Store where, during an hour of encouraging customers to dip lumps of bread into fresh olive oil, we met all manner of lovely people, including Baroness Shephard, erstwhile MP for South West Norfolk, Education Secretary, Employment Secretary and deputy chair of the Conservative Party among many other things. She was a mover and a shaker on Norfolk County Council when I first knew her, when a fledgling reporter knew full well to have his wits about him in the presence of knee-trembling intellect. I would keep checking I'd spelled her name correctly like a nervous passenger repeatedly confirming his passport was in his pocket.

Just time to jog through the market to delicious Strattons Hotel and Restaurant to leave some oil, then back into Norwich once more, this time for a tasting in Jarrolds Deli, to be greeted there by friends and family eager, as ever, to support. Which all sort of dovetailed reasonably well with plans for me to don a penguin suit that evening and attend the old boys' (and old girls') dinner at Norwich School; when grown men regressed, and two headmasters, one past the other present, stood tall, fostered cheers and pride and illustrated greatness with facts both academic and sporting.

Sunday. Time to head for home, but first lunch with Maggie's sister Liz and husband Gary, a one-time drummer, rock band manager and concert promoter, and for a long time the European face of Remo, the greatest name in drum heads and percussion. Just the guy you need when you've just acquired six drums in need of TLC.

A last minute rush to squeeze my mum's old projector, screen and boxes of slides into the car and we were off back to Kent for another sleep over with my Sis before catching an early ferry.

Monday. "Would passengers keep their possessions with them at all times. Do not leave them in the car." Oh, how we laughed. Then, as we were being offloaded at Calais, Maggie realised she hadn't got her mobile. She vanished upstairs again and everyone drove off the car deck. Except me. I was shepherded to the exit but refused to leave without the wife. The memory of her jogging the length of the empty car deck to the cheeky encouragement of the crew is clear as a bell, as is that of her expression when I later found her phone in my coat pocket.

By 6pm we were in the Loire valley, just outside Tours, flaked out in a room at Le Chateau des Ormeaux. Our moment to breathe. Utterly beautiful. After breakfast we bought chateau green tomato and cinnamon jam, and let the resident Labrador lead us on a walk to the chateau's Vouvray vineyard through frosty oak woods dripping with wild clematis. Then it was back on the road through poplar woods adorned with mistletoe to pay a "Surprise! Surprise!" call on old friends, only to find them out. Doh. There was a vague plan to stop somewhere for another night, but the tug of children and home became so strong that on we pressed, trundling up our drive at 4am on the Tuesday morning. Hallelujah. Vital statistics – 2671 miles, plenty of happy customers, too many coffees and grey hairs, one resolution. We won't be doing that again in a hurry.

The dichotomy of cultures that most of the world knows as the one country, Spain, is, for the most part, beyond us. We have only explored Catalonia in part, have dipped a toe into Aragon and spent nine busy days in Cantabria and Asturias. Maggie lived near Seville for a half a year in the early nineties and we had a holiday near Malaga when Ella was a toddler, but we have barely scratched the surface of this complex land. The lure is there, yet Mother's Garden has a firm hold on us, for the already well-documented financial and farm reasons. Spare cash is as scarce as Hemel Hempstead herring fishermen, plus there are precious few people who have a working knowledge of our plumbing and ponies – and sufficient time on their hands – to be able care for the place while we swan off.

I won't bore you with a prolonged travelogue, but here are a couple of anecdotes.

It is so easy to forget the time here, and I mean the hour, the day, the year. The foot of Aragon felt like that in the dreamy heat of summer. People in the squares of hill villages sat and stared, and beneath some ancient arches a little shop stood locked, its opening hours defined as thus:

Abrimos cuando llegamos. Cerramos cuando nos vamos, y si vienes y no estamos es que no coincidimos. We open when we arrive. We close when we go, and if you come and we are not here, it is because we haven't coincided.

Frequently moist and subsequently lush Asturias was our most recent indulgence, a break to visit cave paintings, to find dinosaur tracks and to walk in the Picos de Europa. I could write a book about it. The holiday also coincided with the World Cup Final, when I relaxed a little too much. As Spain lifted the cup, and delirium set in, I was raising a glass of free fizz in a village bar in the mountains. A grandmother ran down the street and into the building waving her arms in unbridled glee before blurring with the general festivities. The few drops of Spanish blood in my veins went immediately to my head and it seemed only appropriate that I burst into song (with suitable apologies to Queen).

"We are the champions. We are the champions!" is what I thought I was singing. "We are the mushrooms. We are the MUSHROOMS!" is what spilled forth in Spanish, repeatedly I am reliably informed. *Campeones* means champions, *champiñones* means mushrooms. Tricky business being spontaneous in foreign languages (I find).

THE GOOD AND THE GREAT

*If, in our world, a kindness makes someone
smile, a comforting word makes someone
feel warm, a hug makes a child feel safe, we
have changed their world, our world.*
Joe and Lorraine Williams

Aid workers Joe and Lorraine, from the charity Imagine, in Mozambique, have been among the most vital visitors to The Garden. People from every continent have swung by over the decade, bringing their stories, cultures and interests to enrich our own. Their great number has included a yoga swami, an Everest base camp guide from Nepal, intrepid south Korean cyclists (yes, all that way), a Dutch cellist, a Columbian sound therapist, an ex-England cricketer, back-packing New Zealand grannies, entomologists from Illinois and young farm helpers from Holland, England, Canada and America.

We absorb them all. The privilege is immeasurable.

And, thank goodness, Joe and Lorraine made it to us. This book is dedicated to them.

Early in 2010, I flew away to Liverpool to speak at Joe's memorial service in the Catholic Cathedral. He had collapsed with a brain haemorrhage while carrying on his and Lorraine's 20-year-long mission to ease suffering among the poor and ill of Mozambique. He was 60. When they laid him to rest in Catembe, where Imagine is based, a great mass of grateful people followed behind the coffin, singing in the rain

172

which was seen as a great blessing. He was a truly special soul. These were my words for the Liverpool celebration of his life.

I've seen the mighty Liverpool play, the famous red, that spirit – just the once though.

It was on a patch of broken concrete and dust, in blistering heat in the welcome shade of trees beside an abandoned building with broken windows. Just down the road from here... in Mozambique.

The ball resembled our torn, worn world, but what it lacked in air was more than made up for by the glee and energy of the young barefoot players, and the happy beseeching and encouragement from the indiscernible touchline, where a boyish coach of about fifty years, grinning from ear to ear, sweat on his pink brow, itched to be tearing about among them.

Just down the road I've been driven past the lip of a vast waste tip, to the lip of understanding, where young children scavenged to survive the day.

I have followed on foot along tracks past sunken shacks where stagnant water moved with mosquito larvae, to higher ground and new beginnings, where women waiting in long line for clean water from new wells sang with gratitude and happiness.

I have seen a priest's unconditional love for Tusha and other infants with no beds on which to sleep, and no chance of a life because of Aids.

I have ridden through the bush on the back of an Imagine pick-up with fourteen others, where all stood to allow more on and everyone weaved themselves together for support.

I have sat on the tiny wooden benches in a child hospice run by Mother Teresa Sisters and listened to life and talk of new plans and what can - will be done.

I have seen a street child with a razor-sharp blade in his pocket melt into smiles on receiving a blanket of his own.

I have met families with new roofs, children with new schools.

Just down the road I have sunk to the earth with the weight of thoughts and watched as two people from this city continued to give completely of themselves, from hugging when hugging means everything, to trying with all their being to help people to help themselves from the shadows of poverty and disease. And, in equal measure, I have seen love.

My guides then and now, to what matters, to eternal faith and optimism, to selflessness, humility, to true fulfilment through compassion for fellow beings are Joe and Lorraine.

Mozambique is indeed just down the road, because whether it was their intention or not, and I suspect not, Joe and Lorraine are the bridge between two deeply distinct worlds, and their lives have and continue to foster vital meaning and understanding, gifting the seed of opportunity to help and, in a way, a chance for fulfilment to so many of us locked in societies that have and know so much, but yet can be so morally and materially distant.

Lorraine – love, thoughts, appreciation, support - long may the softest of voices carry the furthest.

Joe, I will bear all in mind, because, and I can hear you saying how the smallest things can

*change our world for the better, and when faced
with the greatest odds, saying so clearly, "You have
to try, don't you?".*

Finding myself in Mozambique in 2000 was, literally, a watershed. Amid the distressing images of Mozambique's floods in March of that year came a plea for compassion from an English family living there, and they wrote to the Eastern Daily Press, of which I was a deputy editor, saying there was a small charity called Imagine that could put any Norfolk aid to best use.

Readers donated £40,000 in a matter of days. I flew to southern Africa to report on the crisis and how and where the appeal could make a difference by supporting two people who were giving every ounce of themselves to counter extreme poverty, to offer love and comfort to orphans and child Aids suffers, to support hospices, shelters, schools and families. In the wake of the floods, when all of war-torn Mozambique's deep-rooted tragedies were compounded by destruction and water-borne disease, Joe and Lorraine used the money to organise the building of 32 brick houses on safe ground and for the drilling of wells.

Joe, like Lorraine, neither sought nor welcomed recognition. He couldn't see his significance. All experiences and any chance to help were deemed the greatest privilege. There was a boyish love of life, the buoyancy of humour. In just one corner of a country on a continent littered with tragedy they selflessly and with unerring optimism, faith and humility sweated to shed light and hope onto painful truths. It was one of the greatest lessons of my life.

My abiding memory is of their beams and Joe's answer as we drove through the endless shanty town that girdles Maputo, when I asked what exactly we were going to do at the Mother Teresa Sisters' hospice for Aids children that Imagine helped fund.

"Hugging!"

And so we did, without thought of time. Joe and Lorraine were engulfed, then sat upon the ground and opened their arms with unconditional love.

Among the many in Liverpool Catholic Cathedral were three Norfolk people who inspired my journey to southern Africa in 2000. Karen Randell and her late husband John were living with their children Monica and Scott in Maputo when the flooding disaster occurred. It was this family's plea that landed on my desk in the newsroom.

After the service, a woman asked for a copy of my words. Her church supported Imagine and she wanted the congregation to hear something from the service. It was, by chance, St Mary Magdalene's in Enfield, where Maggie attended the Brownies, just down the road from where my late father-in-law David Whitman farmed in north London's green belt. Another wonderful man. Small world.

We adjourned to drink tea and eat cake, then I took my leave to walk, first along Hope Street to the Phil pub opposite the Liverpool Philharmonic Hall, then down Duke Street into the Church Of England Cathedral, and on and on to Albert Dock and the mighty Mersey. Some city it is.

I was staying in a cheap and cheerful hotel close to the airport, where my father was stationed on the guns early in the war before sailing from Liverpool to North Africa with the 1st Army. I sat on the steamed-up top deck of the number 80 bus in and out of the city, through suburbs etched into recent history because of riots. There is deeply depressing talk of Britain being "broken", (Times Populus poll, February 2010) a concept built on the flaws of fractured morals and family values. Spin that theory on to the decay of social cohesion and the quoted figure of forty per cent of the population wanting to emigrate sort of adds up.

I hate this. Britain is currently not my dwelling place for personal reasons, but it remains a beacon of freedom, kindness, charity and community, if only the majority voice of sense and decency, that which has mostly fallen silent under the weight of depressing news, said stop to all the mounting pressure. I've said it before and I repeat it now. Time is what must be retrieved, time for family and others. By living somewhere else you see things perhaps as residents don't. Every time and everywhere I visit I see consideration and courtesy, from the number 80 bus skirting Toxteth to the fellowship of English street markets. The economic schooling of recent decades may be self, self, self, but it hasn't dented our common sense that fellowship is essential to the species. Oh that we can stop judging on possession, park our pride and somehow enrich consciences to keep open minds and hearts, to hope, to keep communicating across the street, across the generations. I actually don't think it is insurmountable because we all sense it is the source of true happiness and fulfilment. Please believe.

Joe and Lorraine Williams, who chose so long ago to go to Mozambique when a bloody civil war was still raging and it was officially the poorest nation on earth, would always remind anyone who would listen that the smallest kind deed can change the world. When I asked Joe how he coped when faced daily with suffering beyond most people's comprehension, he said that if we give of ourselves all is possible.

mother's garden

FOOD GLORIOUS FOOD

If more of us valued food and cheer and song above hoarded gold, it would be a merrier world.

Tolkien

Maybe some of you were drawn to read this book because of one glorious food that is synonymous with the Mediterranean. Fresh olive oil, with all its history and mystery dating back to Biblical times and beyond, is at the very heart of our diet, health and lives here, and for six years it has been a business too.

It may not make our fortune, but we are happy to say that everyone who has bought a container with the Mother's Garden tree on the label, including some outstanding chefs, has bought another and another and another. They, in turn, have told colleagues, neighbours, family and friends, and now we have an online shop. The word is most definitely spreading as we offer the simplest, wisest and most wonderful of foods with the equally simple, essential truths of provenance and freshness.

Extra virgin olive oil is a fruit juice with the water removed. It's not a refined oil. When fresh it has an extraordinary flavour, scent and goodness. If the extra virgin olive oil you use fails the taste or fragrance test then it is most assuredly old oil and also lacking in vitamins and antioxidants that you find in fresh (i.e. most recent harvest) olive oil. People in these mountains who tend the arbequina olive groves wouldn't dream of using anything but the freshest oil, and they are incredulous that anyone should buy any without knowing when it was pressed and bottled. But that is what happens a great deal in northern Europe. Extra Virgin and cold-pressed are understood flags of quality, but age is equally important, as is knowing that the oil comes from a single mill and hasn't been blended. So every bottle of Mother's Garden EV, cold-pressed, single village co-operative olive oil carries the dates of pressing and bottling, which makes us different to the vast majority.

The farm has just one hundred trees, the harvest from which we combine with that of three neighbours to keep our respective families well-oiled throughout the year. This is how the business began. Visitors said they had never tasted such beautifully fresh olive oil and asked if we would send some to England. Impossible, we said. There is just enough for us. Then we thought, and looked around us. Now we work with forty farming families, all members of one village cooperative close to Mother's Garden. We give them fair trade, visit their groves and work in the mill alongside them. Every November, when the finest olives are hand-picked, we sample the oil, and it's carefully stored at the mill until we are ready to bottle and ship it. Our aim is to get the oil from the mill to delis or into private customers kitchens within five weeks, while building "hubs", groups of family and friends, to share a delivery to cut transport impact and costs.

Here endeth the pitch.

Like writing, and the holiday cottage, olive oil is a large piece in the Mother's Garden jigsaw. This has been as great a food adventure as any other, and we endeavour to grow as much as we can on the farm and waste little, and our storeroom is stocked with Maggie's jams, chutneys, pasta sauces, soups, our wine and, of course, olive oil. Delia listed Mother's Garden oil in one of her cookbooks, Maggie posts regular recipes on our website, and we pop up occasionally at farmers' markets and food fairs, peddling fresh olive oil and talking about what we believe in – goodness and taste over appearance, finding every way to cut waste and transport. I was back in the UK just a few weeks ago, visiting my father, and wandering lost among the displays in a modern-day, peel off a bag and help yourself grocers of perfectly-formed produce, looking for green chilli peppers for a stir fry, when I saw a man sniffing a peach. Have you sniffed a fruit lately? Go on, give it a whirl. And, like the gentleman in question, ignore the bemusement of those around you and take a bite. Better to pay for just one unsatisfactory fruit than for a bag full of them. Quite how we have come to judge so much by appearances and not substance (and I'm not just referring to fruit), is something we have to reverse, surely.

   As fledgling businesspeople trying to look tidy, off we have toddled at the onset of our oil trading to the enormous food fair staged in Barcelona, to size up the olive oil world under one roof. It was also an opportunity to visit some friends on their wine stalls in an exhibition hall that could comfortably house a Zeppelin airship, and to try and blag a free lunch. After working out pretty quickly that we could definitely make a go of establishing the Mother's Garden name and our tree logo based on the relatively untapped but obvious market for fresh olive oil, we moved into the main food hall expecting a free feast to cheer us up. I managed to convince a lonely looking Cornish pasty producer on the UK stand to let us do a quality analysis, but that didn't stem the hunger, and all that seemed to be on offer were nibbles. Time rapidly evaporated, so we rushed to catch the bus, cutting through the South American section at a fair old lick, when all at once, I spied a woman laden with a full tray of tomatoes stuffed with soft cheese. I grinned at her, patting my tummy. She raised her eyebrows invitingly, but as I stuffed a whole tomato in my mouth, I noticed her eyebrows were still aloft and her eyes were wide, as if she was waiting for something. Approval maybe. They weren't tomatoes, they were chilli peppers.

Nepalese friend Puskar later got me into chomping on a fresh, small, green chilli with a meal, but the consequences of that food fair greed rolled on for a week.

Born and raised on the north Norfolk coast, I am a connoisseur of superlative sunsets, of the crepuscular light, when time holds its breath between day and night. I have the common contentment of hypnosis by refracted red, when clouds are torched by things beyond sight and reason, above the perfect curve of sea.

As a boy with muddy knees, one sock up, one down, sundown jaw-drops were shared with Leicester Grandpa and Yorkshire Gran on the rough grass of Beeston Bump, a few yards from their home. There, overlooking Sheringham snuggled between the Bump and the hills of Franklin, Morley and Skelding, Grandpa would entreat me to seek twelve-sided thru'penny pieces in rabbit holes: an occasional sixpence too, by golly. I grew up there in so many ways. A few steps further on, towards the putting green and slope to the East Promenade, where my first serious, heartache girlfriend lived, is the bench that bears both my mother's name and the memory of her standing at that very same spot every morning, breathing deeply come sun, hail or horizontal gale.

I can tell you exactly when and where our last five-star Norfolk sunset was before we moved; the second week of January 2001. Blakeney. We stood on the bank, locked together, and wondered what we were leaving. A week later and the same sun was setting on the great Kirby-Whitman relocation to Mother's Garden in Catalonia. Maybe the familiar sky and the constant wonder of it is a vital piece of the puzzle as to why we have stayed for a decade. The last of the light still means the same, and at the end of the valley we are afforded clear air to the western horizon. We have more light, that is true, but you need clouds to reflect the glory. Hence, the frequency of such evening treasure is much the same.

St Martin's day, the 11th of November, is said to herald the *estiuet*– little summer, the bizarrely reliable Indian sojourn before winter. We know it so well. The second weekend of November is when we harvest our olives, and every year we have sweated on at least one of the three days it takes the happy band of friends and neighbours to reap 850 kilos. Dawn at that time of year is different, because at first light, smoke invariably plateaus out at tree tops and hangs in front of the full moon like mist in a Hammer horror film. The fire risk has abated sufficiently for permits to be given out, and farmers with matches start early, clearing prunings and leaves.

In 2009, the St Martin warm spell was ceaseless for about a month, hitting twenty seven degrees one day. The air was as gentle and light as a blown kiss and everyone spoke of the finest *estiuet* in memory. Out came our nets and sun hats, with no Maggie to share the work and words as she was in Norfolk for our first farmers' market and, more importantly, the eightieth birthday of Beryl, her lovely mum, my *sogra* (mother-in-law).

Ella, Joe Joe and I felt the miles, so Ella posted a message to her Nana on YouTube which listed eighty reasons why Nana is Nana. Read them by all means – and see some family photographs – by opening YouTube and searching for nippernana. Back in the olive groves, I and the three farming families with whom we share the task and spoils were joined by my cousin James and friends from Barcelona and Girona. Timelessly we raked, netted and sacked, breaking from work (but not chat) to sit in a circle and log another passing year. We talked of the postponed frost and the enduring colours and contrasts, how the dew could sheen the chocolate bark of young plum trees still wearing a crown of banana-yellow leaves, how types of vines, so difficult to separate in green spring, now each bore a very different shade of autumn. How, for all human distraction, nature and the sun were always unsurpassed, if there was but more time – no, inclination - to be motionless once in a while.

I harvested, chewed the fat and tried to work out in words my heavy heart and incredulity during that Armistice week.

On the one hand there was the annual and wholly appropriate record of ultimate sacrifice during conflicts, in tandem with stark images of coffins from Afghanistan. On the other, someone, somewhere, thought nothing of picking the same week to launch the most popular computer "game" in history, an all too vivid war scenario called Call Of Duty; Modern Warfare 2.

You know the sort of thing – killing without pain or spilling your coffee, only with ever more realistic graphics that the world is fast approaching the time when it will be impossible to tell reality from so-called play. More than 4.7 million copies were sold on the first day alone, raking in £186 million in the US and Britain combined. That's all right then. What a fillip for the economy. It is an 18+ only game, so no chance of youngsters getting their hands on it (oh yeah), while responsible adults will enjoy an "unprecedented level of action as players face off against a new threat dedicated to bringing the world to the brink of collapse" (official product description), before toddling off to bed to dream of camouflage and gun-toting heroics. Then the cynical muse alights upon my shoulder. With such games the so-called first nations are guaranteed the next generation of armed forces recruits.

Were that essential books, like *Vessel Of Sadness* by William Woodruff, sold in such numbers. First published 40 years ago and written by the author of the best-selling autobiography of The Road To Nab End, it is the most vivid and masterful record of wretched war, laid out in absorbing prose that, unlike any video "game", places you at the heart of horror.

Come on, though. Where are we heading?

There was a chilling echo of war here at the same time. They had just extended Joe's village school, sacrificing the pretty almond grove in the heart of the village, a strange and yet pleasing plot hemmed in by houses, to carve out the space for three extra classrooms, play area and garden.

"Look what I found today," Joe said, digging in his pocket. Out came a round of rusty but clearly live ammunition, with a tear in the side and powder dribbling out.

He said he hadn't shown it to the teachers because they would have confiscated it.

I tried not to snatch, then kicked off a short and to the point discussion about what we were dealing with here, before running into the school to alert the head teacher. Joe showed us where he had found it, on a freshly carved earth bank, but there was no sign of any more and the head teacher's interest waned.

"But the International Brigade were stationed right here during the civil war," I said. "It could be a sign that there is a stash of armaments buried nearby."

The council was mentioned, and off I belted. These things were always turning up, I was told.

"But not in school playgrounds! Get a metal detector or something."

I received a "you're not from round these parts, are you?" kind of look, but also a pledge that the mayor would be told.

This tale came spilling out during our chatty olive harvest, when we circled the trees with good friends and the fruit fell around our feet.

That's nothing, someone said. Her partner was the maintenance man in a small village of about a hundred souls four miles up the valley. One old lady showed him into her damp cellar where she'd stored a seventy-year-old mortar and racks of shells, just in case.

I'm never going to be able to tie a ribbon around the great Mother's Garden parcel of memories because it keeps growing and changing shape. The task isn't helped either by constantly wrestling with my own misgivings about life, the universe and everything.

Our 2009 grape harvest has fermented and is now in a five hundred litre stainless steel container that we traded for two lumps of translation work. None of the old oak wine barrels that helpers Harry and Ross battled to clean were pressed into service in the end, because there is always a risk of contamination with old wood, and our expert winemaking friends assured us we were on track for an excellent vintage, if we modernised. Picking, sorting, crushing, punching down, pressing and generally lugging great weights about during the wine making will test your resolve, and had me in the hospital emergency department with chest pains. I'd only just popped into the surgery to say I thought I'd pulled a muscle, but before you could say "A large Jack Daniels please" I was flat out on the couch where a year earlier they'd patched my finger after a lucky escape with a chainsaw, and there were two ambulance men bursting through the door. Turns out, I'd pulled a muscle.

And by the time you are reading this we will know if our highly technical ecological experiment in the vineyard – pruning and then leaving the plants to their own devices – has worked. Well, there are so many other things eating into time we thought it would be worth a shot.

In late 2009 and early 2010, we rekindled a little of the vegetable garden productivity of our prolific early years, with the help of Puskar, who lived with us for a few months. He scattered seed like nobody's business. A sizeable piece of ground by the spring outflow at the back of the house was ploughed, harrowed, rotivated, weeded, fed with pony muck and fenced to knee height to deter boar and tail-chasing puppies. The plot is near the spring and falls on the boars' regular route across our land, so we had to deploy some recycled black plastic and old computer discs to make them stop and think. I know what you are about to say, but I prefer not to dwell on my past failings in the garden defences department. While Puskar and Maggie planted leeks, garlic, broad beans, cabbages, celery, carrots and beetroot, I pottered with iron posts, hammer and old fencing and tried to believe that amid the deep anxiety and depression of economic recession something pretty weird, yet wonderful, might finally be happening in the world.

I keep telling myself, then and now, that there are clear signs of how greed and excess are finally being hauled into the open to be seen for what they are, the roots of moral and social meltdown, the fuel for resentment, anger and violence. People have had enough of being squeezed, pushed and pummelled in an image and work crazed society where profit qualifies everything. We headed for distant hills to try and distance ourselves and our children from rampant commercialism and consumerism, to re-evaluate what was important to us and to try and claim back the one thing you cannot buy, time for family. There is something seriously flawed with a world that puts such a high price on self, while devaluing to the point of destruction those two things that are at the heart of our history and survival. Family and community are the linchpins of civilisation, the essential classrooms of love, fulfilment, communication and compassion, and the armour to protect us from tyrants and tempest. You know it.

It's a truth that is far larger in the Latin conscience than those of Americanized nations. You probably realise, too, that things are changing very fast here in Spain and that something is ebbing away. A 2009 report in Britain, called The Good Childhood Inquiry and commissioned by the charity The Children's Society, says with clarity and sound argument that the aggressive pursuit of personal success by adults is now the greatest threat to children. Something fundamental has to change, it says, both in social attitudes and policies, and it pinpoints family break-up and unprincipled advertising among the factors. The test will be how much that report is still in the public conscience at the time you are reading this, because everything seems so transient these days.

It's not that complicated, is it? When we expose young children to advertising that distorts value and pushes parents into the horrible trap of forsaking time for family in pursuit of yet more money, so that they can somehow pay for it all, then we fail. When people are so channelled to concentrate on possessions and appearances that they fall into the abyss of envy, stress and depression then families fail. And witnessing all of this, believing that is how it must be, are the next generation.

Spain's recession and the well-documented hand-ringing in the Eurozone have hit us, of course, and I have had to dabble with a little English teaching once more to try and make ends meet. I'd taught a group of adults in the village for a couple of years early on, which was more valuable for the contacts and friendships that I forged. The last class petered out one hot August, when we wrapped up the course book and talked about our favourite television programmes. One was Fawlty Towers. I wondered how much genius was lost in the dubbing.

"Really?" I said. "You weren't offended by the Manuel character?"

"Why?" said Oriol. "He's very funny, no?"

"Yes, but he's from Barcelona."

"No, he isn't. He's from Mexico."

I had neither plans nor ambition to teach again (I find it utterly exhausting) but I fell into it, and the little reward has helped. Not that I have managed to improve anyone's English very much. Somehow I found myself elected on to the parents' group at Ella's senior school, and in the blink of an eye someone came up with the novel idea that I could offer English revision classes for students who were struggling. It would have been churlish to say no because there was little else I was good for, and it would save me from other committee work. For an academic year I attempted to break through to fifteen deeply depressed children who were forced to watch their friends go home then sit with me for an hour from 5.15pm. I tried repeatedly to suggest a more positive time for these classes, but the powers that be deemed it impossible. In the end (don't tell the parents), I managed to ignite a spark of interest and impart some words and grammar by forgetting their course work completely and doing tricks, playing card and board games and running pop and news quizzes.

However, volunteering to teach didn't save me from other responsibilities. Parents' groups here have to organise all school books and run the canteen. I don't mean wear hairnets and slop peas onto plates, but be responsible for all details including the monitors and finances, and joining the queue occasionally to test the menu. I'm always up for a free meal, but going line by line through legally verbose Catalan catering contracts made me feel like a student forced to attend an after school Martin Kirby English revision class.

mother's garden

COUNT THE PETALS OF
THE MOON DAISY

*Writing is the only profession where no one
considers you ridiculous if you earn no
money.*

Jules Renard

Funny old carry on, this "career" thingy, don't you think?
Why are you doing what you are doing? More to the point,
what would you rather be doing, or have done?

I was a journalist for twenty two fizzy years (which
makes me a master of nothing) and have had four books
printed with my name on the cover, but when people ask
what I do for a living, the title "writer" is still itchy. Maybe
that's because I don't come up for a public airing very
frequently, and have never been adept at selling anything, let
alone myself, which is most definitely part of the game these
days. Or, maybe it's because I'm just an old hack who likes
nothing better than the solitude of thought, a keyboard and a
few words to play with.

Despite what Jules Renard says, I had it drilled into
me during formative years by a mother (who tried to make
me wear a cravat to her lunch parties) that income is society's
step-ladder to the lofty pigeonholes of success. She was most
definitely a matriarch of the shiny shoes variety, a stickler for

189

appearances who wasn't going to leave anything to chance where my future was concerned. Why didn't I rebel? Oh, I did, when she wasn't looking. Because, because, because. Patricia Grace Armson was one hell of a woman, a lovable, potent, potpourri of beauty, iron will, pride, energy and gross maternal aspirations, a devoted mother who got up at 5am to prepare breakfast for visitors in her large guesthouse, and then worked in an estate agents during the day before cooking an evening meal for up to twenty people to somehow pay for the education that would turn me into someone, er, important. Despite repeated doubts on my part, and that of a host of educators, she never stopped believing.

That's why, despite pubescent ponderings of literary fame and some seriously dodgy attempts at poetry, I parked the whole idea of trying to pen a masterpiece aged just 17, and was on track to explore English literature and the female students' block at Leeds University, when I spied an intriguing job ad.

Actually, it was a tad more dramatic than that. I'd come home from school early in my A level year, after what should have been a day of rigorous career contemplation and advice, to be greeted by my mum at the door of the kitchen holding a laden picnic hamper, china and all.

"How'd it go, dear?" she beamed, probably expecting the words scientist, astronaut, doctor or civil servant to spill from my lips. She was right on one count.

"They think I'd make a good social worker."

Mother froze, but the hamper didn't. Even when it hit the floor with the sound of greenhouse explosion she didn't blink. Her reaction couldn't have been worse if I'd said I'd been spying for the Soviets. To her it meant several things; my appearance would relax somewhat (to a comfort zone I have so happily gravitated to anyway), I would most probably join the Labour party and a ghastly union, and I'd have to take on one of the toughest, most thankless tasks in society. She wanted a first class ticket on the gravy train and a title akin to nobility for me, not a coal shovel on the footplate.

I'll never know if I had what it takes to be a social worker. Probably not. It's up there with teaching. You need to be exceptional, selfless, dedicated and with unerring fortitude that you can carry responsibility and evoke positive change while being subjected to unbelievable, unremitting scrutiny.

So, I joined a profession that did the scrutinising, which was far, far easier.

I'd backed out of the room as my mum mumbled something about dead bodies as she dropped shards of china back into the hamper, and had gone into the living room to re-evaluate my situation. I'd sort of bought the idea of social work, but with that out of the window I picked up the newspaper and scanned the jobs section, trying to see if anything suited. It wasn't a big advert, but it leapt out. Trainee journalists wanted. College places funded AND a salary. I could be paid to learn to write.

Before my mother had fully recovered the power of speech I was in front of her again.

"I've had another idea," I said. She sank into a chair wondering what was coming next.

"I want to be a newspaper editor." Note how I pitched that. Her wide eyes computed the information and a faint smile finally flickered across her twitchy lips. I held up the advert and said, with a pinch of truth, it was what I really wanted to do, but didn't, up until that point, know where to begin.

Twenty two years later, when the wits on which I'd lived were wilting a little, I resigned from the same newspaper company that had saved that day back in 1977, and moved to Catalonia.

It had been a considerable privilege working all that time for Eastern Counties Newspapers, now known as Archant, not least because of some of the exceptional people I worked with and encountered while trying to measure up as a local newspaper journalist. My colleagues numbered a few Norfolkmen and women, but the majority were from away, from Yorkshire, Cheshire, West Bromwich, London. The potency and variety of that life were the keys. You never really knew anything – what the day would bring, what hour it would end, what weight your words would carry (or not), how far adrenalin and judgement could be sustained by caffeine and Mars bars, how unfeeling you could become to other people's catastrophes. That last point still worries me about society in general and adrenalin-rush journalists in particular. You are trained to shed timidity, sharpen tenacity, nail the English language, law, local government and shorthand, take anything in your stride, and stay at your post until the job is done, which can be as long as a piece of string. And at the end of the day, when I was an editor, and because of the endless information conveyor-belting past me, I felt I knew too much. It's a moot point, but I now wonder if England, maybe the world, suffers from bleak news overload.

But, I am still and always will be a newspaperman, and am pained to see them, especially those giving their all for local weeklies and daily titles, fighting to stay in the public conscience, while frequently wafer-thin, multi-channel television fills every second to overflowing, turning part of the brain to idle.

We don't have a television any more, although I am inclined to go online to enjoy the sometimes brilliant, occasionally hilariously burlesque, frequently puffed up London media circus of print and broadcast, where some highly entertaining egos have soapboxes and loudhailers of varying sizes. Vital stuff, of course, from an industry that, despite spawning my favourite radio programme (The News Quiz), is so all powerful, confident and vociferous that I'm sure it would eat itself if there was nothing else to get its teeth into. Being an ex-journo, I suppose I read a lot between the lines and have more than an inkling about the lifestyles, privileges, and the fizz of adrenalin from the cocktail of notoriety, fat pay cheque, and freedom these people enjoy. Being a journalist is such a buzz. Being paid a packet to speak your mind or pick someone else's has to be one of the best gigs going, even if the taxman monitors expenses more closely these days. Such unfettered freedom brings enthralling petulance too. A particular recent favourite was the vacuum between the last general election and the announcement of the Tory-Lib Dem coalition, when for several unbearable days of minutes and pages to fill, frantic journalists lobbed all their toys out of the pram because they wanted a potty (something to go on), and they wanted it now.

I said London not national media because it so often seems that distant from greater Britain. Long live the regional press, say I, living and working in the communities they serve, with just a receptionist between them and the people they are writing about, which rather helps to keep things in perspective. That is the journalism I have known. And I am glad to still be writing for the Yorkshire Post and Eastern Daily Press, albeit just offering up a monthly minestrone of tales from the farm, much like this book.

We call this, incidentally, the Whitman farm, in memory of Maggie's dad David, and also because Maggie, a registered *agricultura,* calls the shots. As you have probably gleaned by now, trying to keep on top of ten organic acres – No, Bob Dylan, I ain't going to quit – is a maddening puzzle of tears and joy, which we juggle with giving the children as much of ourselves as we can. Which is why the writing must come before dawn, usually from 5.00am until 7.30m, before other issues cloud my mind. Then I take tea upstairs, pull on Maggie's red wellies and do the rounds of the chickens, cat, dogs and ponies. Sometimes Maggie is about too, but such is the length of her day, what with all of the above, olive oil and the holiday cottage, I try not to stir her when I get up. And I am happy with this rhythm for now, sort of knowing that as the children grow and we face up to the fact that we can't physically cope with so much land my time for writing will grow, if the grey matter holds out.

This will be my third book in the ten years we have been at Mother's Garden. *No Going Back – Journey to Mother's Garden* was the first, but it was never intended to be. It came out of necessity, riding on the back of the two Channel Four documentaries and was picked up by publishers Little

Brown, bringing in a few valuable sovereigns. But the firm intention on coming here had been to devote what writing time I could find to completing *Count The Petals Of The Moon Daisy*.

That novel has proved to be an extraordinary journey in itself. I'd started it just before we left England, inspired by nature and music, but I'd needed space to breathe to make it work. It took seven years in the end, and was published in 2007, not that that was the end of the story. As I write the fourth and final draft of screenplay is finished and being read by funders, producers, directors and actors. Eye Film, the production company who bought the rights, are trying to pull together the finance and the team to make Moon Daisy into a feature film.

Perhaps the film will have a different title to the book. So much is in the melting pot right now.

The story is of how an American woman, a violin virtuoso whose world is falling apart, is forced to make a life-changing journey to the lost realm of the English water gypsies, to the heart of the big-sky wilderness of the Norfolk Broads. It's a mystery of roots and old wisdom, of truths, of the healing beauty of music and nature when two women's lives, separated by a century and an ocean, weave closer and closer until they touch. It is a story of loss and re-discovery that places Norfolk, England, at the heart of American history and touches the roots of hundreds of thousands, possibly millions of English-Americans. It tells of the lost and little known folk who lived and worked on the wild wetland, and it weaves natural beauty with the English pastoral music Herbert Howells. I hope, too, that it questions head on the wisdom of modern values, as a great number of people are beginning to do.

Write about what you know, they always say. Despite my disparate roots, I grew up in Norfolk and learned to sail and lose myself on the Broads. Maggie's background is classical music, and through her I have a deep appreciation for early Twentieth Century English composition that entwines with my love of natural history writing from that era. Weaving these together in a novel became an essential endeavour for me, and I'm eternally grateful for the support of my family and a great many others for helping me to fulfil it. Now, I am learning the art of writing screenplay, and will be a producer on the film project, if we get to production. Whatever happens, it has been so enlightening and rewarding working with Frank Prendergast and Charlie Gauvain and of Eye Film, who I would never have met if it had not been for my old friend and former newspaper reporter turned television editor, Mike Talbot.

Just maybe, if we get to film, I will really start to earn some money from writing and we can stop trying to juggle so much.

It crossed my mind to give this chapter the title STAYING AFLOAT. Income, or rather the alarming lack of it, is for a great many people what often pulls the plug on such grand adventures in sunnier climes because, unless you have a lump sum on which to draw for evermore (like incompetent executives with gross pay offs), the grim truth is the challenge of finding work that pays a reasonable wage can be a cloud that will follow you around dampening spirits until you give up.

We have met a fair number of people who have sold up in England and have been living here on the proceeds and their pensions. But even that can prove stressful or impossible if, as came to pass, the exchange rate collapses. But the bleakest stories, many with sorry endings, are of families like us, who move to the Mediterranean nations with heaps of belief and determination, but no spare cash, and who try to string out an existence on what work can be found. That is not easy at the best of times, but lob in a world recession and it becomes nigh on impossible.

We are still going for several reasons, not least luck. Somehow, we have pieced together a jigsaw of little incomes that promote one another, rooted like our Mother's Garden tree logo on the cover of this book in our fortune at finding this farm.

The newspaper writing has been constant, I am glad to say, and this work has lead to new friendships and brought a host of people here. I also write a monthly column for Catalonia Today, a small but growing, erudite magazine which is published in English for a predominantly Catalan readership. My relatively short, sharp lumps of vitriol, entitled *Heading For The Hills*, allow me to let rip on all manner of social and global issues that have come into sharp focus since I changed my altitude and perspective. Gawd knows what they make of me, but I persevere, encouraging this Latin state to learn from the harsh lessons of more Americanised nations where rampant commercialism has placed an unbearable load on family and sanity. Blow a large raspberry, I tell them, to all that they know in their heart of hearts to be utter rubbish.

Pthzzthzzthzzp! to the cult of celebrity and the triffid seeds it plants in young people's minds about image over substance, possession over knowledge, self before others.

Pthzzthzzthzzp! to the devaluing of the invaluable, like sitting down together as a family to eat, talk, laugh and clear the air, far from the drivel of so much TV; like using all our senses by walking out into the real world that isn't, strangely enough, scentless, unfeeling, boxed in glass or wrapped in plastic; like learning to communicate – you know, having the security, confidence, vocabulary and the capacity to listen, simple yet essential skill that unlocks our greatest needs of support, appreciation, understanding, humour, discovery, resolution and love.

Don't get me started. Again.

I endeavour to dabble in irony and frivolity sometimes, and the piece about Mother's Garden extreme sports - washing the cat, nude blackberrying, taking off overalls in clean kitchen without first removing muddy boots, that sort of thing – raised titters and eyebrows. But mostly I'm ranting and waving the rather torn standard of common sense on global issues that scare the hell out of me, like nuclear power and GM food.

Were that I could make a living out of writing. Not so, so far. Hence we diversify, and Mother's Garden obliges.

For the first few years we got by on precious little. It was just about possible because we did not have a mortgage and few lumpy overheads. I was bringing in a few sovereigns teaching English and writing articles for any publication in the world who would pay. We were also regularly distracted by the filming of the first and second Channel Four documentaries about us that had kicked off six months before we left the UK, for which we were not paid, I hasten to add. Our riches were chiefly an overpowering sense of adventure and bare-foot contact with nature. We grew copious amounts of vegetables and fruit and sold some to an organic shop. We locked into the neighbourly practice of sharing rather than paying for labour and machinery. But we were treading water, unable to significantly improve our draughty and dusty home where we cooked on an old gas ring or the open fire.

Then two things happened. Mum passed away and Joe Joe developed some serious health issues. We needed to re-evaluate a great many things, and begin to think about earning some serious money.

It was about this time that our niece Rosa, with a little help from Ella, drew a picture of a tree of many fruits. It was to become our Mother's Garden trademark as we began selling fresh extra virgin olive oil in the UK, while planning how we could accommodate the people from around the world who wanted a flavour of this land, this life.

While trying to get olive oil business off the ground we decided to tackle the ruin of a house that sat next to the track to the farmhouse. It was a mess, home to rodents, a deceased Seat car and all manner of debris, but it was also a gift. It was an established dwelling. Getting rural planning permission for new homes here is impossible, and I've already said the nod and a wink days from blind eye village mayors are long gone. Take the risk and you face paying to have your home demolished, at your expense.

But, our wreck was on local planning maps - invaluably legal. All we had to do was restore it, if (underlined) we were prepared to swallow dive back into mortgage debt, and also fight our way through Spanish red tape to get it listed as a holiday house business. Deep breath, pinch your nose time. Did we want to risk getting out of our depth? We weighed the pros and cons, and what tipped the balance was that it was so rare to have an existing property. So we went for it, telling ourselves it would pay us back one day.

The six winter and spring months of rebuilding were hectic yet so fulfilling, with two guys on site every wintry dawn bar one, when it was snowing so heavily nobody was going anywhere. Top man was David (PIM PAM POM), he of Herculean strength, ready smile and, vitally, an artistic bent. He and second builder Marc welcomed our hands-on approach, didn't flinch if we veered off the straight lines of the architects drawings, and proved loveable, tireless antidotes to the old adage about dodgy Spanish builders. All the same I thought, David had lost the plot when it came to the staircase.

Back then, when the appropriate Catalan words deserted me, I'd become quite adept at the Mediterranean parlance of facial contortions, animal noises and arm waving, but even this seemed woefully inadequate. "You are kidding me", I gestured with a grimace, slap of the forehead, half-stifled schoolboy giggle and pointed finger. There was no way, José, that anyone could build an unsupported staircase with a handful of thin bricks and some rapid cement.

The flight to the first floor in our old farmhouse had me baffled from the first day I clapped eyes on it. Heavy steps have been loaded onto a wafer thin arc of bricks that rises and tucks over like the trail of a Space Shuttle rocketing into orbit. There was no way in a month of Sundays something so apparently gravity-defying could be so solid and remain standing for two hundred years or more without the aid of copious amounts of iron bars embedded into the wall.

So, when David said he was willing and eager to put in a traditional Catalan staircase in the holiday cottage, I

198

plonked myself down on the floor to watch. Step one was to get a long plank positioned on its side at the correct angle and far enough away from the old wall to allow for the width of the staircase. Fair enough. I was with him so far. Next, with pencil, cardboard and nails David marked out and fixed a narrow, curved template to the inside of the plank, just wide enough to support the edge of the outside bricks. Then, with his fellow builder Marc mixing cement and handing him bricks as fast he could, David proceeded to build the arc.

I lost the plot at this point. The result was very pretty, but one nudge and it would surely collapse like me after a large Spanish brandy. The next morning he laid another tier of thin bricks on top, at right angles to the first ones, but which still only made the arc an inch and a half thick at most. These were left for a while to set before on went the brick steps that collectively must have weighed a ton. There were no iron bars, no reinforced steel, not even interlocking bricks.

How on earth...? It's down to the angle of the arc at the top, it seems, with the curve sending all the weight down at an angle. Incredible.

Up rose the walls, along came a truck carrying the massive pine roof beams, and we stood on the meadow and watched as the house was topped out, David having used his shoulder to move the house-length ridge timber into place. Two Moroccans, and Khalid, joined the crew to help with the plastering and rendering while a digger carved a hole for the pool, and Biba our increasingly overweight hound got terrible wind and went off her food. We are talking flower-wilting, paint stripping intensity, yet she seemed particularly well and content. Delirious more like. All the guys on the building site were feeding her, and the root of the foul odour was Hammadi who, once the weather had improved, had started cooking spicy stews on an open fire.

Maggie blubbed and I got some dust in my eye the day they finished, their last task being to carve their names atop the sundial David had built on the south facing wall above the front door. But there was no time to dither. Our first booking was in the April and we only had a couple of weeks to paint the walls, linseed the windows and doors and furnish an entire house. The last days, hours, and minutes were what you might call a nip-and-tuck, skin-of-your-teeth, no-time-for-a-shave-let-alone-a-haircut, I'm-never-doing-this-again-as-long-as-I-live experience.

With just three days to go there we were, the last in the Carrefour hypermarket queue at closing time at 10pm on a Wednesday night, with a cot mattress, several bags of kitchen utensils and two store staff propping up a massive pile of seven pieces of rather classy 'on special offer' cane furniture including a sofa. There was fat chance we were going to get it all in or on top of battered old Range Rover, but we had no option but to block that thought out of our tired minds. We were out of time. Going home and pacing up and down praying for Carrefour to deliver everything before Saturday, just wasn't an option.

I smiled weakly at the woman on the till, who clearly wanted to get home and wasn't expecting to look up and find someone furnishing a house. There was nothing for it but to start giggling. We've all been there, haven't we, when the cocktail of adrenalin, embarrassment and fatigue only needs a *soupçon* of the sense of the absurd and you're off. I said to Maggie that if worse came to worse, we'd just have to pile the furniture six feet high on the car like the Hillbillies and take the back lanes home at twenty miles an hour under the cover of darkness.

Then, just when we needed a miracle, there was Marc, builder number two, minding his own business with his fiancée Marta at the next till buying a few things for a light supper. Bless them, they saved the day, packing their car with one of our new chairs, several bags and the baby mattress despite the risk of raised eyebrows back in our town where they were getting married in the following September.

Good. We finally had something for our guests to sit on; bed bases too, accumulated over the past months along with kitchen equipment, lamps, towels, bedside tables and other bits and bobs. There was still, however, the small matter of the big mattresses and the dining room table and chairs.

Only one thing for it - a frantic Friday night foray to Ikea in Barcelona, the night before we opened for business.

All giant stores do my head in and promptly make me forget why I went there in the first place. When Maggie and I first got together in 1992, she was taken aback by the forest of Dettox bottles under my kitchen sink. The explanation was simple. Every time I went for an occasional big shop, I would slide into a trance, like all the wide-eyed blokes you see drifting through the aisles on their own. I always knew I had to remember Dettox. It always leapt out at me when I saw it on the shelf, but not remembering whether it was a case of I definitely didn't or did need it, I always bought one just in case.

Clever the way these establishments rig the lighting and play music to befuddle the male mind, and then channel you past everything knowing there's a fair chance you're going to have more than one of those "oh, I hadn't thought of that" moments. It's no coincidence why supermarket shoppers are made to walk miles through a maze of marketing to get to the daily essentials, like bread.

I went into a daze, but Maggie stuck to the mission and tracked down some fantastic cut-price deals on some really high quality wooden dining furniture, including a 12-seater beech-block table and wooden high-chair, and in no time at all we had another mountain to move and had parted company with circa a thousand Euros. That would have panicked me if there'd been time for it to sink in, but the lights were going out in the underground car park and we'd over-looked the fact that the low ceiling and exit meant we couldn't strap things on top of the car this time.

All we could do was wedge everything into the car, and I had to drive with my arms through the legs of the high chair and up and out into the dense Barcelona traffic where we had to pull over and do the Hillbillies thing all over again.

As I stopped at the end of the drive the following afternoon and, as agreed, sounded the car horn to let Maggie know I was back from the airport with my old friend Mike and his family (our first guests) Maggie was washing the bottom step of the outside stairs. I'd already taken the long way back, detouring into the nearby town on an unwanted sightseeing trip, and there was no way I could stall any more. Maggie emerged into the twilight through the front doorway, resplendent in her dungarees with a mop and bucket in hand and managing, despite the rigours of the day which had started at 6am with the building of the Ikea dining table and twelve chairs, a glowing smile.

I want to trebly record she's a ruddy marvel, adorable and how I love her beyond the stars. Her eagerness and ability to make everyone who comes to the farm feel welcome is well known by all who understand her, but it has been tested to the limit over the years, and sometimes beyond. The holiday house only happened because of her determination, and the beauty of it is down to her inspiration and perspiration. "That," she would say, pointing at the old ruin, "has got to happen. We need it. We need to get on with it. Now would be good."

People who come to see us now have a beautiful house with pool, loaded with the charm and features Maggie always envisaged, but that is not the point. It is a business that rewards our hospitality, another valuable string to our bow. And we no longer have to necessarily share our house or put ourselves, or friends and family who come to stay for that matter, through the stress of living together under one roof. You've all, no doubt, had people to stay. Wonderful isn't it? But, however much you want to see someone it can be a strain and a bucket load of work. Well, in our first year here in Catalonia, we had seventy; singularly adorable, accumulatively sapping to such a level that you wanted to scream.

So, there was Maggie, proudly guiding Jane and Neil, Mike and Kirby (her real Christian name) through the open kitchen and dining and living space, three bedrooms, shower room and bathroom of *La Vinya Del Pare*. Our first reward was to be told by Jane, whom we'd never met before, that she thought it was so beautiful she wanted to cry. And she and Neil sent us a lovely message after their week with us saying it was one of the best holidays they had ever had.

I must tell you, of another visitor to *La Vinya del Pare* that Easter in 2005. The butterflies here are plentiful and varied, but are sometimes out-classed by the moths. The house had only been finished for a couple of days when, on the old red stone beside the French doors, someone noticed something. A huge peacock moth, Europe's largest, had come to call and seemed happy to hang around. We saw it as a blessing.

Six years on, our lives have been enriched by visitors from around the world, who have come for the wildlife, the wine, the wilderness, the walking or just the simple pleasure of unplugging themselves from their normal existence for a week or two.

All the same, given the number of plates now spinning above our heads and an eagerness not to be tied from Easter to October, we are thinking of ringing the changes. Maybe the cottage and a lump of the land will be sold, maybe we will seek people to share the labour and our dream of it being some sort of a healing/time out retreat. For we are agreed, the writing and the olive oil business might in the coming years be more than enough to keep us on our ageing toes.

mother's garden

NOW YOU?

*With the possible exception of the equator,
everything begins somewhere.*

C.S.Lewis

It's July 2, 2010, and I am two months and two days beyond my book deadline.

No doubt the volcanic chaos will be a distant memory by the time you are reading this, along with the Clegg effect in the pantomime of the UK election, and England's dismal showing in the World Cup. But the significance and consequence of all three, along with my water problems, have been immeasurable here of late, and I have been much distracted. Anyway, where publishing is concerned an act of God is the trump card of excuses.

But it's true. When Mount Eyjafjallajoekull blew its top in Iceland Ella was almost stranded with school friends in Romania, and Maggie caught one of the last flights out of Stansted. Plus the goings-on in Downing Street and the hyperventilating media have been just too bizarre and entertaining not to follow, while watching England's abject failure to live up to expectations in South Africa was a case of exhausting masochism. As for my waterworks, well...

During a two-week lull in the holiday cottage business we plunged into the pool almost hourly and attempted to steady the heart-rate and brainwaves in readiness for our few days away in Cantabria and Asturias on Spain's northern coast. Cousin James was here to hold the fort and handle any foal arrivals, so I tried to sort a few last minute tasks to ease his burden, like topping up the pool for example. Ella stopped swimming and stood, as always, with the pipe above her head, enjoying the flow; until it started spewing mud. She screamed. I screamed. I ran up the land to the distant well like my arse was alight, despite the ninety five degrees of heat.

The well couldn't have run dry, surely? That would be a catastrophe. The farmhouse, the holiday cottage and the land relied completely upon its unwavering good nature. But there could be no other reason for dirty water. We must be down to the last dribble. I felt like Robert Green after the ball trickled over the line.

Then, as I loped, gasping, the last few metres through the top vineyard I could hear gushing. Water was pouring out of a blown pipe that must have been softened and weakened by the blazing sun. Our pump, fitted a couple of years ago, is ludicrously powerful and could service a small town. We were pumping mud because the soil around the well was saturated and this was leeching back into the shaft. Phew, I thought. We are not out of water. It could be far worse. But, back at the farm buildings, the glutinous mud had jammed open an electrical tap feed to the main water tank, which overflowed, flooding the barn and, for good measure, blowing up the hot and cold water pump. Yet another act of God. Anyone would think she or he didn't want us to have a holiday, or for this book to be published.

Out of my office window the dagger leaves of the irises have grown to swords and the flag flowers have come and gone. A myriad of flower essences float among the birdsong. The fig trees shine green with life and lower their boughs to the ground with the weight of fruit. Cicadas dominate the dusk and our new young cockerels claim the dawn.

Oh, I'm forgetting another excuse. In April it rained and rained. Great dollops of H2O fell at the rate of two inches a day. A dampness chilled the house and we were out of dry wood save unserviceable furniture. The ponies were miserable, Jess the cat was holed up in the barn and we were holed up in our kitchen, burning the last of uncle Philip's childhood chest of drawers that we brought with us but which fell apart soon after and has littered the yard ever since. We should have been doing VAT returns and trying to navigate the good ship Mother's Garden through the economic storm, but we were heady and tired and there was marmalade to make.

Firewood won't be a problem next winter.

Forest clearance after the tree-breaking snowfalls was in full swing across the sierra for a considerable time, hence the *grrr* of saws from all quarters, and we too cut and burned to hopefully spare us a summer torching. On the new clear days wooding was such satisfying spring labour because the mist of dew that painted the valley with milky brushstrokes extinguished all fear of accidental fire storm, and amid the damaged trees you could see where you have carved air. Better still, lily of the valley, bee-orchids and wild sweet peas, dormant for goodness knows how long, had stirred from beneath the blanket of needles where the pines once blocked light and life. With Maggie feeding the fires I had no choice but to get stuck in, so whistled happily to take my mind off writing guilt, and when the damp-down deadline of noon arrived I was invariably humming, so I jump into the shower with bramble inscriptions on my forearms and forehead.

We are, perchance, the most far-flung members of the Norfolk organic growers group. Mother's Garden is tended with the benefit of all manner of gleaned wisdom here and there, most recently when Maggie attended a meeting with mum Beryl and sister Liz, to hear James Dexter of Earwig Organics expound on something we feel strongly about regarding the future of mankind and that sort of thing. Dust. Not the stuff on your bookshelves or under the bed, that minestrone of human skin particles, plant pollen, specks of hairs and textiles that has made vacuum cleaners a multi-billion euro business, but rock dust, akin to what Eyjafjallajoekull so effectively used to bring Europe to its knees.

No, I'm not advocating regular volcanic chaos, but there are other ways.

Rock dust remains a mystery to many, yet it's potentially the planet's way of saving us from ourselves, if only we would wake up to it and take off the blinkers of commercialism. During ice ages, glaciers pulverise, grind up and move rocks over vast distances. This dust re-mineralises, acting as a host to micro-organisms that enrich soil and hence sustain our whole precious environment that we are doing so much to trash. The problem is the last ice age melted away 10,000 years ago and now most of the soil has been drained of the broad spectrum of minerals needed to maintain its health. The crucial significance? It has been proved that plants grown in soil re-mineralized with rock dust are so resilient, they don't need any chemical concoctions – and production and goodness can rocket. Regardless of how appetising fruits and vegetables may appear in the shops, vitamin and mineral content has dropped significantly in the past forty years, in many cases by fifty per cent. This is because the ground itself is losing its goodness. While we all wait for the dust to settle, ponder on your core health issues hopes, our collective crucial need for prevention rather than cure, and how the rock on which we live could fundamentally address some of civilization's greatest ills.

Intrigued? Look up rock dust on the internet on sites like www.seercentre.org.uk. Any good garden centre should be able to advise. Also dig into books like *We want real food* by Graham Harvey. Test rock dust in your window box, or back garden, and slowly but surely the world may get better.

Having three vegetable patches dotted about presents inevitable irrigation issues. Resourcefulness is watered by necessity and stubbornness. We flatly refuse to throw away anything remotely recyclable, until every last alternative use can be explored. Sometimes, the consequences are crude. Sometimes, they are quite charming. Puskar suddenly decided with Nepalese vigour to dig up the small terrace in front of the house, where Maggie and her watering can had cultivated copious quantities of vegetables in our first two years. Certainly the soil is fertile, but we had moved our patch because we bought chickens, and their enclosure and afternoon free range territory was there.

We're going to need fencing and water, I said, leaning on the low wall. Then my eyes wandered to the wine making overspill stacked against the barn, including old oak barrels that could no longer be trusted for our moonshine, but are way too charming to part with.

Puskar, meanwhile, began constructing a barricade of cane, hazel and old string that has a certain something too. And it did, for a short while, deter the brood from scratching out our rocket, tomato, pumpkin, pepper and melon plants.

This was during the second week of April, and Puskar broke off his labours to wish us "Navavarsha", Happy New Year.

"Get a grip," I said.

"No, now is our Nepalese new year. It starts very nicely now."

"Your 2010 starts in April."

"No. The year is 2067." A flash of white teeth. So there you have it, in case you were worried about predictions of doom in 2012 just move to Nepal and bypass Nostradamus.

I mentioned our new cockerels. They are my sixth and final excuse. While Cameron and Clegg were constructing their nuts and bolts alliance I was knocking together a hen house, made entirely from (I preen) 100 per cent recycled materials, including pallet nails. It took eight pallets in all, snaffled on three raids to the dump, but there's no blueprint I'm afraid. It sort of evolved, with Joe designing a neat door that turns into a flight of chicken stairs. The project was urgent because the day old chicks we had billeted in the barn under a heat lamp at Easter had filled out faster than journalists at a free buffet.

The good/bad news is that there appear to be seven, possibly eight, cockerels. Great for Sunday lunches, if I can do it. Should repeatedly muttering the mantra of food provenance fail, I will ask my mate Jaume or maybe his mother to assist. The women round here don't seem to blink an eye.

I've also sought to ease the barn overcrowding by catching and releasing (several kilometres away in the forest) the fattest couple of rats I've ever seen either here or back in the UK. They outwitted me for days, lumbering between the various animal feed stores. We are always happy to have the swallows and bats back, but the rats were seriously ugly and potential pullet slayers.

Are you up for all of that? Do you want to shake the tree of life?

The degree to which your feet itch will depend a great deal on your age, family, aspirations, rural or urban preferences and current degree of contentment. Much that has been written or broadcast about "life-change" doesn't challenge the comfort zone, it reinforces it. Take it easy, soak up the sun, get off the carousel. Spain is still, despite the various headlines of late, a prime location for attempted opt out. Fair enough, but I wanted to be honest with you.

What I read in week-old British papers mailed to us by family, always sends me up the garden for a few deep breaths. I'm supposed to be keeping in touch with what's going on back home, but, without fail, the stories reinforce my sense of being further and further out of orbit. The world is doing its thing and I'm floating in space, trying to make sense of a whole load of weird stuff like, for instance, why so many people would prefer to live somewhere else. I keep telling myself it can't be true. Not that many people can be as nutty as us. Yet, every time I am back in the UK someone invariably starts a conversation about exodus and how they dream, either idly or in glorious Technicolor with Dolby Surround Sound, of joining it; about brain-frazzling stress and a burning desire to jack it all in; about feeling squeezed to the point of losing the plot; of (as I said at the outset of this book) wanting more colour and contrast in their lives.

"POP POP POP - There just ain't enough bleedin' bubble wrap to go round mate," a London cabbie wryly remarked after he'd kicked off a chat by asking where I lived, as we pulled up behind several Mercedes owners who could afford the congestion charge. "The world may be round, but I wake up in a box, eat breakfast out of a box, watch the world going mad on a box, then drive around in a box. Others, they work in a box and stare at a box all day. Bleedin' loopy."

So, what is going on?

One indisputable truth is that the once laughable notion of selling up and shipping out has been sown and watered in millions of minds over the past few years and it has grown like Jack's beanstalk, promising golden days and an end to the incessant grind and grey and increasingly grim nightly news. It is seriously alarming how many once immovable Brits pipe up that they are worried silly about, well, you name it – the cost of living, crime, overcrowding. I get the impression that they are tossing and turning in bed because the voices in their heads are having a ding dong like this.

"Enough, ENOUGH. Go on, get the hell out of here. Everyone else is. You saw the headlines in the Sunday papers."

"Hang on, hang on. Don't be daft. Let's not do anything rash. Think of the kids, the family, friends. Of course we can't leave. It's an absurd notion. Plan a holiday. Order a takeaway and take our minds off it with a DVD like we normally do."

"Rubbish. Think of the money we could get for this place. Better still, think of the sunshine. No more battling into work: no more battling when we get there and no more insane office politics. We could swap the semi for somewhere beautiful, and the family can come to us. Think about it. How much time do we spend with them now, anyway? All we seem to do is work, sleep or sit in traffic watching wiper blades."

"Don't be ridiculous. Do NOT rock the boat. This conversation is seriously dangerous."

"This conversation is about life. What do you want to do with the rest of it? Where do we see ourselves in two years or five years? The world is changing and we can change too. We really can afford that villa we saw in the paper, the one with the balcony dribbling bougainvillea down to the eternity pool. It is not a dream. And we can still watch sport on Sky."

"HOLD IT. Hold your horses. See sense. What about our friends?"

"Exactly. Two couples have gone already – one to Oz the other to Bulgaria."

"OK then, what about … what about our pension? What about learning a new language? What if we get sick? What if we don't like it? What then, smarty pants?"

"So, going to give up the dream just like that just because of a few challenges? Hundreds of thousands have worked all that out and are doing it and so can we. We need a change. We're more than ready for a new challenge. And what is the alternative? Carry on plodding along? We have said we are sick of the struggle, sick of the costs, sick of the weather, sick of chasing our own tail. Let's do something about it. We have worked long and hard enough, haven't we? Stop day-dreaming about it and start living the bloody dream. I dare."

All the same, it dried my throat to hear Britons are still buying one-way bus/plane/ferry/train tickets at the rate of well over one hundred and thirty thousand a year, in spite of the recession. If that many are doing it, how many are thinking about doing it? A pre-general election poll for The Times stated bleakly that people were deeply pessimistic about the state of Britain, believing that society was broken and heading in the wrong direction. Nearly three fifths of voters said that they hardly recognised the country they are living in, while a staggering forty two per cent said they would emigrate if they could.

Hang on a second. How else are people likely to feel given all they are seeing and hearing? The 21st century world is turning into one enormous revolving door, and everyone in chop-chop-work-work-busy-busy Blighty is being shown pictures of how easypeasy lemon squeezy it is to trade a modest home in Hampshire, Hemel Hempstead, Halifax or Hartlepool for a little piece of Mediterranean, Australasian or North American paradise that will turn the neighbours back home green. They are going because they are told continuously that they can; which is stressful whether you are plucking up courage to go, or just watching over the garden fence as the neighbour loads a removal truck bound for Brittany. And nobody likes being left behind or missing out. I have always suffered a queasy feeling of doubt and sadness, maybe even envy, when someone leaves. I can't fully explain why, but it has been with me through life, from when friends moved schools, colleagues handed in their notice, or nice people in the street sold up. So goodness knows what the other half of the country, those who wouldn't dream of changing, must be feeling right now. To counter this, there was a powerful and uplifting comment posted by a reader on the Times website about the *Is Britain Broken?* story.

"Here's a crazy idea," he wrote. "Instead of moaning about it why don't you do something? Volunteer at a local charity, clean up your street, see a bit of litter? Pick it up, etc etc. One decides one's own fate. If you work hard you'll be fine. If you sit back and think the world is rubbish then it will be rubbish. Emigrating changes the venue but it rarely solves the problem, if you emigrate because you're unhappy it won't fix it. Anyone I know who is happy after emigrating did so for a career opportunity, out of need to survive, to be with someone they loved or after working hard all their lives for a dream retirement. I don't think Britain is broken. Having seen a third world country up close I think Britain is brilliant."

Maybe the exodus from Britain is, fundamentally, as it was for us in January 2001, about time, or rather the chronic lack of it. You know – you can have anything you want, if you work hard and pay up. But, the real bill is time for yourself and your family.

It might be that thirty or forty or fiftysomethings are getting an irrepressible itch to experience more in this life while still able. Or, perhaps some adventurous happy people have pinched their noses and launched themselves off the high board despite their apparent riches because they too buy the adage that the greatest regrets in life are not the things you did, but the things you didn't do. Fulfilment, or rather a yearning for it, can be rocket fuel.

It's a riddle not so easily solved.

Every time I am back in lovely Britain (that is genuinely how I feel) I get an increasingly unnerving feeling that something is going skew-whiff. Me, possibly; or maybe not. I have struggled with my distance from the organised world, namely an unease that can so easily slide into despair at the rise in human habits of control and non cooperation, leaning so heavily on dogma, polarisation, self, claiming the moral high ground or crying victim as a means of shedding responsibility. Taking the easy option of putting on the armour of orthodoxy when it becomes nigh on impossible to say and mean things like "I am listening", "I was wrong", "I am sorry". Human frailties of greed, suspicion, prejudice and fear have become brittle to the point of fracturing, fostering aggression, mistrust and anxiety.

Politics is a fine example. I have never found it positive, accessible, or able to simply show in word or deed what I see as core values – constraint, tolerance, compromise, clarity and honesty. Government by conflict is so depressing – two rows of pinstripe clever trousers arguing the toss relentlessly, wrapped up in the sixth form debating society challenge of winning, rather than solving a problem. Maybe that might change now.

What about toil, personal responsibility, charity and hope? That goes for a great lump of the media too, now so obsessively negative, forever fanning political disagreement into conflict, feeding fears and prejudices, never able to take back what in haste they have shouted from the rooftops.

There could be many reasons that fuel any thought of leaving.

I worry, too, about the extraordinary financial pressure people are under these days, so compounded by recession. The world's economy is a super-glue that holds a lot of things together, sure. Work, reward, so work harder and have more. But stay the whip a second. Some people are now so transfixed by the debt juggernaut in their rear view mirror that the journey through life must be seriously scary. Stopping isn't an option. The only lines that are ever drawn are performance targets that are always higher year on year.

Or maybe it's the 24-7 tirade of news, views, gossip, digital flat-screen 100-channel celebrity-nuts entertainment that fries the brain to the desensitizing point of not knowing what is real any more. Have you noticed how fast images change in adverts these days? And what is this fixation with image and celebrity?

While clever bunnies in marketing turn idle curiosities into wants, and then wants into needs, everyone is weaved into a surreal existence of endless material aspirations and the sweat of paying for it all. Meanwhile, mindless, unfulfilling armchair entertainment panders to the voyeur in all of us, killing conversation, when the best things of all, free things at that, diminish absurdly - fresh air, community and the best show on earth, Mother Nature. Hearing about the number of children with computer games and televisions in their bedrooms drained the blood from my brain, just as lying on a bed watching something or other until you nod off must do to theirs. Where young minds are concerned, it is so worrying how precious little is left to the imagination anymore, because there's money to be made.

And the word moderation is becoming redundant. People are schooled to consume more and more, while they, in turn, are consumed. Rabid commercialism and a chronic obsession with celebrity, and the control they seem to bear, are, on my bleakest days, as mesmerizing as the vortex of a plughole.

I'm not finished yet. There is another trend I fret about. Some people call them targets, like the sort you shoot at, only this is mathematics by which all are compared, contrasted and judged. And I mean all, from mental health cases to sales of toilet paper. You can measure toilet paper. It's uniform and comes in a tidy roll. People with mental health issues don't. There is no way of defining targets where need and wellbeing are so diverse, yet society, it seems, demands results for absolutely everything. That, frankly, is not possible, so all manner of clever ways are devised to redecorate the statistics, from air-brushing to wallpapering in a world where human capacity for humility, compassion and honesty is crushed by fear – fear of being judged as failing, fear of losing a job, fear of not getting re-elected. Presentation is king; substance is the knave.

And as for the gluttony of those who morally must know what they are paid by their corporations or clubs is indefensible, I shudder. Bankruptcy is not a word confined to finance.

It's seven years since my first "heading for the hills" book *No Going Back – Journey to Mother's Garden* was published, and ten since the Kirbillies loaded their truck to the gunnels with two little children, two Springer spaniels all manner of debris and headed for Spain. We copped out and, touch wood and whistle, things have worked out, but not without serious pain and doubt. I have long lost count of the days when we nearly gave up.

For all the stats on mass emigration I'd be curious to know how many people give this kind of existence a whirl and then return home. We have seen it many times. Sometimes Shangri-La, so far from familiar people and things, is not all it is cranked up to be.

And while I'm at it, can we talk a bit about glass houses? There is a palpable and understandable disquiet about the influx of immigrants into the UK, about the strain on the system and that integration is not happening fast enough if at all. I am an immigrant, sort of. I hedge, because I'm a Briton living in southern Europe, like millions of other north Europeans, with a British passport in a drawer somewhere, trying my level best to grasp the lingo and integrate, but with no firm plans to forsake my nationality. Why? Well would you? I may return to England one day, probably the cul-de-sac called Norfolk where I was born. Or we may move somewhere else on the planet because our list of worldly ambitions grows not shrinks. My nationality is, I suppose, my identity badge, and while I remain a wanderer it serves some sort of purpose in a world that persists in being paranoid about pigeonholing, stereotypes and differences. But the more I meet people of different races and creeds all I see are similarities.

If someone had told me ten years ago I'd be on first name terms with Catalan, Spanish, Moroccan, Cuban, Algerian, Ecuadorian, Colombian, Czech, Slovak, German, Dutch, Belgian, French, Polish, Romanian, Russian, Lithuanian and Argentinean people I would have scoffed. But here we all are, cheek by jowl for whatever reason. And the Catalans, like the Spanish in general, are, on the whole, remarkably open and tolerant. In fact I am constantly amazed to see how patient they are when having to deal with foreigners who persist in expecting them to understand their language, like, er... er... the British.

What is good about this rural Latin life?

We're together. We have barely missed a second of Ella and Joe Joe's childhood. I like to think it has in some ways proved far easier to parent here in time-travel Spain, primarily because neither of us is absent.

We like it how there is so much more emphasis on the family, more awareness of one's responsibility as well as rights, and less peer pressure among a young generation who are far better guided in family values and social awareness. This might sound like small beer, but believe me it isn't. You can go out and eat anywhere and children are openly expected and accepted, but there is always a firm hand too. The boundaries of behaviour are far better understood and so the freedoms are greater, the inter-generation frictions seem much lower. Certainly where we are in the sticks this has a great deal to do with generations staying closer to home, believing in sitting down and eating together as a family, children growing up with grandparents and great-grandparents, being tactile, knowing why and when it is important to use touch. Of course, so-called modern pressures, attitudes and anxieties are coming increasingly to bear here, and just how rapidly the fast-food, eat-on-the-run, multi-channel TV Americanization grows is worrying to witness.

Notwithstanding the single thumping negative of the distance between ourselves and our families and friends, we see our experience as a privilege. This may sound sickly sweet, but it is the simple truth – here we have space, beauty and the rare blessing of being around for each other and our children a great deal of the time.

But – and make that a big fat but – this isn't Britain. It works differently, talks differently. Language, communication in all aspects of life, is a monumental issue. Come a crisis in the doctor's surgery, not understanding and not being able to explain could at least be distressing, at worst a matter of life or death. Like immigrants everywhere, learning the language of the land is a justifiable obligation, a social necessity. Meanwhile, for the pre-retirement masses like us, there is the small matter of staying afloat. Although I wish it did, money does not grow like olives. So we are our own bosses, trying to make ends meet by running a small holiday house, exporting extra virgin olive oil, farming, writing and walking the fine financial line like most other people. We bob, weave, creak and frequently spin just as we did in Britain, and we still haven't fully learned one lesson, that we are human beings and there is nothing shameful about just *being* once in a while. We still haven't got the hang of the vital but increasingly de-valued life skill of doing absolutely nothing occasionally.

Got the picture? If you haven't by now I can't help you.

So there you have it, the second dollop of life down on the farm we call Mother's Garden.

Enough said. I'm off up the land, to take writer and countryman Adrian Bell's advice. After talk, walk; that's essential, he wrote. After dispersing yourself, go and look for yourself. I might play Lou Reed on my ipod and sing as gustily as my mum did to Sing Something Simple, or prop myself against an olive trunk and listen to the birds. For sure, I will think about shaking a tree or two. It is the almond harvest season, after all is said and done...

BIBLIOGRAPHY –

Just a few books to water the mind

Adrian Bell – **Men And The Fields** (Little Toller Books. ISBN-13: 978-0956254528)

Adrian Bell – **Corduroy** (Penguin. ISBN-13: 978-0140290707)

Lilias Rider Haggard – **A Norfolk Notebook** (Faber&Faber, search in second-hand bookshops)

William Woodruff – **Vessel Of Sadness** (Abacus. ISBN-13: 978-0349118116)

Donald Culross Peattie – **The Road Of A Naturalist** (GK Hall. ISBN-13: 978-0839828907)

Graham Harvey – **We Want Real Food** (Constable. ISBN-13: 978-1845292676)

Richard Wilkinson, Kate Pickett – **The Spirit Level** (Penguin. ISBN 13: 978-0141032368)

Anna Quindlen – **A Short Guide to a Happy Life** (Random House USA. ISBN 13: 978-0375504617)

WEBSITES

To find out about fresh extra virgin olive oil, follow our blog and/or to stay at Mother's Garden, see
<u>www.mothersgarden.org</u>

Other related websites

www.imaginemozambique.org

www.eyefilmandtv.co.uk

www.EDP24.co.uk

www.yorkshirepost.co.uk

www.iberianature.com

www.tartdesigns.com

www.heathertamplin.co.uk

www.seercentre.org.uk

www.earwig-organics.co.uk

www.gardenorganic.org.uk

www.biodiversityislife.net

www.slowfood.org.uk

COUNT THE PETALS
OF THE MOON DAISY

By Martin Kirby

(Novel, Pegasus ISBN 9781903490297)

'*Inspirational on many levels, not least of which is going with the urge to follow our own dreams .... It is informative and entertaining and to be recommended as a summer holiday read*' Norfolk Magazine

'*Impressive debut novel....A rare treat in these sour, self-centred, cynical and celebrity-infested times*' Eastern Daily Press

'*A gripping, compelling and beautiful tale. It has great depth and wisdom even for those who already know the truth of the message; that to restore the worn and troubled spirit it is necessary to actually engage with wildlife.*'
Harnser Magazine

Moon Daisy spans the Atlantic and the years, a story of roots and ghosts, music and nature.
Violin virtuoso Jessica Healey sits in her London flat thinking of killing herself, when the phone rings and there begins a rare and beautiful English journey.
Through a 19th century orphan's journal she finds herself carried to a lost world of water gypsies and teeming wildlife.
The Norfolk wind blows to her soul, and the secret of her very being is revealed as their two lives, separated by a centure, weave close and closer ....until they touch.